William J Johnston

Telegraphic Tales and Telegraphic History

A popular account of the electric telegraph, its uses, extent and outgrowths

William J Johnston

Telegraphic Tales and Telegraphic History
A popular account of the electric telegraph, its uses, extent and outgrowths

ISBN/EAN: 9783337090289

Printed in Europe, USA, Canada, Australia, Japan

Cover: Foto ©ninafisch / pixelio.de

More available books at **www.hansebooks.com**

TELEGRAPHIC TALES

AND

TELEGRAPHIC HISTORY.

A POPULAR ACCOUNT OF THE ELECTRIC TELEGRAPH—ITS

USES, EXTENT AND OUTGROWTHS.

By W. J. JOHNSTON,

Editor of "The Operator."

New York:
W. J. JOHNSTON, Publisher,
No. 9 Murray Street.

PREFACE.

Some time ago the subscriber published a number of anecdotes relating to telegraphy, which were received with an unexpected degree of favor. They were so extensively copied in the newspapers as to set him thinking that the preparation of such a book as this would please the reading public, as well as members of the telegraphic profession. Hence the undertaking herewith put on the book market as a candidate for popular favor. No more is claimed for it than that it presents, in a methodized and compact form, a comprehensive summary of such telegraphic information as is likely to be valued by the general public, and of use to the operator because of the convenient method of its presentation—varied, as is desirable, with lighter matter. Very considerable labor has been expended upon it, in the hope and belief that it will occupy an unique place among those books which instruct without being tedious and entertain wholesomely. Should this expectation be verified, the subscriber will be justified in his confidence that the reading public and the profession will in a new instance show their appreciation of that sort of literary work which constructs miscellaneous materials into an edifice not wanting, as he trusts, in symmetry and beauty. The well-read operator may find individual passages herein which he has met with before; but it is believed that he will be the readiest to appreciate the judgment and industry which have put them exactly in their proper places as portions of a book.

The subscriber's modesty would lead him to claim even less than he does for this his latest publication, had he been solely engaged in its production. He will add no more than his hearty acknowledgments of the valuable assistance rendered him by Mr. Henry G. Taylor, a New York journalist whose experience and graceful ease of expression give him distinction under the severe test of metropolitan competition.

W. J. JOHNSTON.

CONTENTS.

PRE-ELECTRIC TELEGRAPHS.. 7
Signaling among the ancients — Telephonic system of the African negroes—Signaling by sound in Montenegro—Fire communication in war and otherwise—Dr. Hooke's telegraph—The semaphore—Semaphoric blunder and its result—The word "telegraph"—Prediction quoted by Addison.

THE ELECTRIC TELEGRAPH—ITS BEGINNING AND DEVELOPMENT. 14
First lightning-rod man—Frictional electricity discovered—The Leyden jar—Experiments to Franklin's time—His famous kite experiment—Robert Stephenson's boyish imitation—Lomond's electric signals—Lesage's invention of electric telegraph, using twenty-four wires—Reiser's thirty-six wire telegraph—Succeeding experiments to Morse and subsequently to present time.

INTRODUCTION OF THE ELECTRIC TELEGRAPH IN THE U. S........ 29
First American line—Apathy of scientists, press and public—Why the *Herald* refused to encourage the telegraph—Cornell and Morse—First apparatus—Interesting relic—First week of telegraph—Slender returns—Humors of early-day telegraphy—Countryman, turkeys and telegraph—Mr. Stearns and obstreperous church bell—Honor to whom honor is due — Ronalds — Morse—Henry—Vail—Claim for laborers—"Be jabers, who dug the post holes?"

A CHAPTER ABOUT OPERATORS AND MESSENGERS.................. 50
The operators' view of human nature—Their faithfulness—Their literature—Their difficulties and trials—Epileptic telegrapher—Armless operator—Deaf operator receiving by sound—The "lightning striker's" blunder and a case of jealousy—Recognizing by touch—Love over the wire—Love disappointment in humorous verse—First marriage by telegraph—Absconding operator caught by novice—Wonderful speed in telegraphing—Messenger service—District telegraph boys and the various duties they perform—Anecdotes of them — Uniformed messenger mistaken for policeman.

THE TELEGRAPH IN WAR.. 70
Earliest military signaling — Introduction of field telegraphy—Field telegraphy described—Dangers to apparatus—Firing guns by electricity—Telegraph in civil war—Its great value—What General Sherman said of it—Origin of U. S. Military Telegraph—Cost of service during war—Duties of cipher operators—Official acknowledgment of their services—Anecdotes of military operators' ready wit, heroic courage and nervousness—Funny war story—Another—Military operators' poor quarters—A provident telegrapher—Richmond taken—Receipt of the great news—Lincoln's assassination—Grand feat of Prussian soldier, and heroism of French female operator.

CABLE TELEGRAPHS.. 95
General—The Atlantic cable—First suggestion of it—Its origin—Organization of company—Laying cable—The Great Eastern—Discouragements—First message—Suggestor wittily silenced by Mr. Field—Cost of first Atlantic cable—Recent improvements in cable laying—Mr. Field's services—Cable operators—Cable codes—A specimen—Its interpretation.

HUMORS OF THE TELEGRAPH... 107
Economical Irishman—Timid old lady—Apprehensive Texan—Witty, incongruous and rhyming telegrams—A "killing" blunder—The "additional wurred"—A furious message—Satchel by telegraph—Snubbing a king—A proper old lady—Little "Johnny Russell"—Peter to Margaret Flagarty—He couldn't be fooled—"She writes like a man"—Model (?) Maine man—Hollow and "hello"—Fooling savages—"Onnateral fixins"—Chicago and ———— —Witty illustration — Electrifying loafers — Shocking the negroes—Blindfolding the "masheen"—A crammer.

CONTENTS.

TELEGRAPHIC "BULLS" 123
A fatal "bull"—Matrimony killed by a "bull"—Instances of operators' "bulls"—A lord's mistake—John Brown and Seaton Bros.—Ale or oil—Too much coffee—Blessings in disguise—A profitable mistake—A military "bull" that was not all a "bull"—Senders' "bulls"—Habit and halibut—"Bulls" from bad spelling—A fishy story—Tragic "bull"—Injustice to operators.

LIGHTNING FREAKS AND TRAGEDIES 138
Deaths from lightning—Effects in different countries—A triple tragedy—Curious freaks of lightning—Some wonderful instances—Lightning in telegraph offices—Operators killed.

SHARP PRACTICE BY TELEGRAPH 143
Abuse of General McClellan's name—A modern "St. John"—Big swindle in Toledo—"Spiritualistic" swindling—Rappers' tricks—Their magnets—How to make them—Sir Charles Wheatstone's experiments—Two good stories of sharp practice by operators—Tampering with cipher message—The biters bit—Great bank swindle—Barb's telegraphic trap for burglars.

THE TELEGRAPH AN UNIVERSAL INSTITUTION 160
A well-traveled message—Spanish peasants and telegraph—Telegraph in Morocco—China—India—The East in general—Japan—In Africa.

THE WEATHER REPORTS 168
Death of General Meyer—His account of storm signal system—Its value to commerce and agriculture—The New York station—Cipher reports of weather—Difficulties of signal service—Early opposition—Origin of weather reports in the United States—Smithsonian Institution—Professor Henry.

THE RAILROAD TELEGRAPHIC SYSTEM 179
Originated in England—First instance of train dispatching in this country—System at Grand Central depot—Moving trains by telegraphic orders—Official instructions—Train dispatchers and operators — Their responsibility — Thrilling incident — Operator who "forgot"—Noble operator—Latest inventions in railroad signaling—The train telegraph—No more screaming engines—Supplying locomotives with water by electricity—Fun on the railroad—Waking the Pullman porter—Operators' anti-suporific.

ELECTRICITY AND LIFE 202
General remarks—Electric girl of La Perriere—Electrical lady of Nevada City—Electricity on dinner table—Feeling pulse by telegraph—Development of growth by electricity—Uses in surgery and dentistry—Electricity as a healer—An "anti-fat" story.

OUTGROWTHS OF THE TELEGRAPH 218
The electric light—Edison's description—The light at Niagara—Experiments in San Francisco—Proposed illumination of Holyoke—Use in stores, steamships, and in war—The telephone—How constructed—The German name for it—Its invention—Telephone service meter—Transmitting sermons by telephone—First instance—Mr. Beecher's—The telephone in Jersey City law courts—Communicating between ships—Use in wooing—In military operations—Music—Humors of the telephone—The singing telephone—Yarn from Pine Bluff—Joke on reporters—One for Dawdles—Marriage by telephone—Telegraphing by light—The photophone—Electrical egg hatching and seed germination—Theatrical thunder—Toothache cured by electricity—Gas lighting and bell ringing by same means—Electricity as an umbrella—In taming horses—In connection with Moody and Sankey's meetings—Telegraphing by electrical air currents—Maps by telegraph—Magnetic magic writing—Electric driving power—Electricity in managing refractory horses—Engraving by electricity—Diagrams of targets over the wire—Electric combs and brushes—New uses for the sun's rays—The ocean a source of electricity—Suggested use of electricity in executing criminals—Slaughtering cattle and killing whales by electricity—Electric clocks that require no winding—Telegraphing by steam at sea—Electricity in steam—The Edison electric locomotive—Description of it—Electricity aiding weary cash girls—Conclusion.

TELEGRAPHIC TALES

AND

Telegraphic History.

PRE-ELECTRIC TELEGRAPHS.

When signaling as a mode of communication was first adopted, no amount of research can ascertain. We find it difficult to conceive of a time when it was not convenient, if not necessary, and when human ingenuity was incapable of providing it.

One of the earliest recorded systems of telegraphy for signaling over long distances originated among the African negroes, and has been practiced from time immemorial. The means used are telephonic, the signals being read by sound, and not by the eye.

The "elliembic," as the instrument used is termed, is still in existence, and used in the Cameroons Country, on the west coast of Africa. By the sounds produced on striking it, the natives carry on conversation with great rapidity, and at several miles distance. The noises are made to produce a perfect and distinct language, as intelligible to the operator as that uttered by the human voice.

It is hardly necessary to add that the existence of this contrivance, capable of such useful effects, implies evolution, probably carried on through a series of ages, from devices which, we may presume, originated in the very infancy of human society.

One of these still prevails in Montenegro, where, when a shepherd in the mountains finds himself in want of society, he sends out at random a peculiar kind of yell, with a view of attracting the attention of any one similarly situated, who may chance to be within hearing upon some other mountain side, and may also feel a desire for conversation. It is well known at what a great distance shrill sounds may be distinctly heard in the mountainous regions. The unseen friend, whose ears have caught the sound, responds in the same way, and then begins a dialogue about their flocks and herds, or any other country gossip; and should there chance to be news of public interest, such as of any important person or foreigner passing that way, the receiver of the intelligence shouts it out in the open air for the benefit of the mountain nearest to him, and so it passes from one to another through a considerable part of the country.

In saying that signaling by sound probably anticipated all other methods of telegraphing, we are simply saying that the most natural and obvious mode of communication, namely, that by means of the voice, was the first made of service in the rapid transmission of intelligence over long distances.

The employment of objects to be seen was a later expression of human ingenuity, intended to better answer the demand for easy and far-reaching communication. And what better for this purpose than fire—a ready servant and the most available for its conspicuousness; real even in the glare of day, and made intense by surrounding darkness during the night?

Accordingly, we find records of the use of fire-signaling during the Greek and Roman wars; and in the writings of Polybius, about two hundred and sixty years before Christ, there is an account of a signal corps attached to the military. Down through the ages "fire-swingers" were employed as signal men.

It is related that at the siege of Vienna, John Smith, the explorer of Virginia, used the plan of Polybius with effect, to arrange with the besieged forces for a sortie, he having learned it from the Turks.

The quaint old English works of 1650, or thereabouts, tell of "a marvelous device by which those who know may converse so far as light may be known from darkness." As a matter of course, every reader is acquainted with the modern use of the fiery cross, and certainly with the telegraphic use of fireworks.

In 1684 Dr. Hooke proposed a kind of mechanical telegraph, which, however, was not carried into operation. He prepared as many different shaped figures in wood, as, for example, squares, triangles, circles, etc., as there are letters in the alphabet. He exhibited them successively in the required order, from behind a screen, and proposed that torches or other lights, combined in different arrangements, should supply their place at night. Twenty years later Amontous, of Paris, exhibited some experiments before the royal family of France and the members of the Academy of Science, showing the practicability of the system.

Semaphoric signaling contrivances were in use in

various countries down to within a half century of the present year (1880). That employed by the English Admiralty was not abolished until the end of the year 1847.

In contrast with the convenience of the electric telegraph it was cumbrous and costly. The expense of working and mounting the line from London to Portsmouth was three thousand three hundred pounds ($16,500) per annum.

Though of great service to the government, it was, of course, only available in clear weather. Vexatious interruptions continually took place, and droll accidents occasionally resulted from the sudden cessation of communication, from a fog, or similar cause, during the transmission of a message.

When, for example, the British army was fighting under Wellington in Spain, news was anxiously expected from that great commander through the Admiralty signals. The public was in a feverish excitement, when one day the disastrous message was received: "Wellington defeated."

The funds were violently agitated, the people and the government were bewildered, and terrible rumors of enormous slaughter and great loss of guns, colors, and ammunition were heard on all sides. It turned out, however, that, just as the word "defeated" had been deciphered at some part of the line, a sudden mist had come on and cut off the remainder of the message. When this inopportune visitor had passed away, the public mind was instantly relieved with the news that the message was not "Wellington defeated," but "Wellington defeated *the French*."

Lest readers should take exception to the use of the word "telegraph," with reference to signaling before the introduction of the electric telegraph, it is interesting to know that in an article published in "Nicholson's Journal of Philosophy" for October, 1798, and entitled, "An Essay on the Art of Conveying Secret and Swift Intelligence," by Richard Lovell Edgeworth, the word "telegraph" is frequently used, and in such a way as to show that it was then a common current term. The following extract from the paper shows what could even then be done in the way of instantaneous transmission of intelligence: "In September, 1796, the lord lieutenant ordered me to prepare telegraphs for an experiment before his excellency. In consequence I constructed four new telegraphs. I had found that the large machines, thirty feet high, with which my sons talked, in September, 1794, across the Channel, between Ireland and Scotland, were liable to accidents in stormy weather, etc."

In the grand march of human progress all previous methods of distant communication were surpassed in general availability by the electric telegraph, which, associated with locomotion by the agency of steam, introduced a new era into the history of civilization.

Very curiously, Addison, in No. 241 of the *Spectator*, December 6th, 1711, quoting from a mediæval writer of monkish Latin, realizes the instrument used for telegraphic purposes in this nineteenth century.

He says:

"Strada, in one of his Prolusions, gives an account of a chimerical correspondence between two friends, by the help of a certain loadstone, which had such virtue in it, that if it touched two several needles, when one of the needles so touched began to move, the other, though at never so great a distance, moved at the same time and in the same manner. He tells us that the two friends, being each of them possessed of one of these needles, made a kind of dial plate, inscribing it with the four-and-twenty letters, in the same manner as the hours of the day are marked upon the ordinary dial-plate. They then fixed one of the needles on each of these plates in such a manner that it could be moved round without impediment so as to touch any of the four-and-twenty letters. Upon their separating from one another into distant countries, they agreed to withdraw themselves punctually into their closets at a certain hour of the day, and to converse with one another by means of this invention. Accordingly, when they were some hundred miles asunder, each of them shut himself up in his closet at the time appointed, and immediately cast his eye upon his dial-plate. If he had a mind to write anything to his friend, he directed his needle to every letter that formed the words which he had occasion for, making a little pause at the end of every word or sentence, to avoid confusion. The friend, in the meanwhile, saw his own sympathetic needle moving of itself to every letter which that of his correspondent pointed at. By this means they talked together across a whole continent, and conveyed their thoughts to one another in an instant over cities or mountains, seas or deserts. * * * * * If ever this invention should be revived or put into practice, I would propose that upon the lovers' dial-plate there should be written not only the four-and-twenty letters, but several entire words which have always a place in passionate epistles, as 'Flames, Darts, Die, Languish, Absence, Cupid, Heart, Eyes, Hang, Drown, and the like.' This would very much abridge the lover's pains in this way of writing a letter, as it would enable him to express the most useful and significant words with a single touch of the needle."

We must now take our readers from the elegant periods of Addison into an account of the successive experiments and discoveries which led up to the invention of the electric telegraph, and afterward to its remarkable perfection as we now know it.

THE ELECTRIC TELEGRAPH—ITS BEGINNING AND DEVELOPMENT.

"Coming events," says a time-honored maxim, "cast their shadows before." Long ere the electric telegraph became an "institution," its feasibility had been anticipated by scientific minds, with greater or less clearness. There is nothing more interesting in the history of experiment than the successive results of the attempt to dominate the electric fluid. The imagination is awed by the sublimity of human endeavor, which, in their turn, overcomes one natural force after another. Water, air, fire, steam, lightning, have had to succumb to the potent spirit of man, "for whom all things were made;" and the future will, doubtless, see still stronger instances than the past, of the triumph which intellect and courage gain in the struggle with nature. Probably never is this glorious victory due to one man. The electric telegraph is no exception to the general rule. As Robert Sabine says: "It grew up little by little, each inventor adding his little to advance it toward perfection."

BEFORE 1794.

Our familiar friend the lightning rod was an appliance of the earliest civilization, namely, that of the ancient Egyptians. These people, alas! had experiences of the persistency of the brazen-cheeked lightning-rod man, who fitly represented the first and simplest process in the subjugation of the electric fluid.

We are in the dark as to the history of man's relations with this subtle agent from the last days of ancient Egypt until about six hundred years before Christ, when Thales, of Miletz, discovered that the rubbing of amber (electron, in Greek) produced what is, perhaps somewhat clumsily, called frictional electricity. Two hundred years later, Plato attempted the first theory of electricity. Ten years before the Christian era, Plutarchus is recorded as having described the electric phenomena observed in his time. Through a very long hiatus we arrive at the date 1690, A. D., when Otto Van Guericke, of Germany, made a friction electric machine. Thirty-eight years afterward, Etienne Grey, of England, discovered the difference between conductors and insulators; and, in the following year, he and another Englishman, named Wheeler, succeeded in transmitting an electric shock through several hundred feet of wire. The Leyden jar was invented in 1745, by Musschenbrook, of Leyden, Holland. It may be described as a glass jar or bottle used to accumulate electricity. The jar is coated with tin foil within and without nearly to its top, and is surmounted by a brass knob for the purpose of charging it with electricity.

We next turn to our own shores and the experiment of the illustrious Franklin, who gives the following account of it, in a letter written by himself to Peter Collinson, Esq., F. R. S., London, which probably contains about all that is definitely know in relation to the American philosopher's discovery of the analogy between the electric spark and lightning. This had been previously conjectured. So early as 1708 Dr. Wall had pointed out a resemblance between them. In 1735 Grey, whom we mentioned just now, had stated that they differ only in degree; and four years before Franklin's great experiment, the Abbe Nollet gave more substantial reasons than had been adduced by Grey, for agreement with him. But to Franklin's letter, which is taken from a quarto volume published in London in 1774, and entitled, "Experiments and Observations on Electricity, made at Philadelphia, in America, by Benjamin Franklin, LL. D. and F. R. S."

FRANKLIN'S OWN ACCOUNT OF HIS EXPERIMENT WITH THE KITE.

"As frequent mention is made in public papers from Europe of the success of the Philadelphia experiment for drawing the electric fire from clouds by means of pointed rods of iron erected on high buildings, etc., it may be agreeable to the curious to be informed that the same experiment has succeeded in Philadelphia, though made in a different and more easy manner, which is as follows:

"Make a small cross of two light strips of cedar, the arms so long as to reach to the corners of a large thin silk handkerchief when extended; tie the corners of the handkerchief to the extremities of the cross, so you have the body of a kite, which being properly accomodated with a tail, loop and string, will rise in the air like those made of paper; but this being of silk is better fitted to bear the wet and wind of a thundergust without tearing. To the top of the upright stick of the cross is to be fixed a very sharp pointed wire, rising a foot or more above the wood. To the end of the twine, next the hand, is to be tied a silk ribbon, and where the silk and twine join a key may be fastened. This kite is to be raised when a thundergust appears to be coming

on, and the person who holds the string must stand within a door or window, or under some cover, so that the silk ribbon may not be wet, and care must be taken that the twine does not touch the frame of the door or window. As soon as any of the thunder clouds come over the kite, the pointed wire will draw the electric fire from them, and the kite, with all the twine will be electrified, and the loose filaments of the twine will stand out every way, and be attracted by an approaching finger. And when the rain has wet the kite and twine, so that it can conduct the electric fire freely, you will find it stream out plentifully from the key on the approach of your knuckle. At this key the phial (Leyden jar) may be charged; and from the electric fire thus obtained spirits may be kindled, and all the other electric experiments be formed which are usually done by the help of a rubbed glass globe or tube, and thereby the sameness of the electric matter with that of lightning completely demonstrated. B. FRANKLIN
"*Oct.* 19, 1752."

Right here is the proper place to record an amusing anecdote of Robert Stephenson, who worthily bore the name which his father had made immortal, and exemplified in his fondness as a boy for scientific experiments that "The child is father of the man." This young gentleman, we are told, was very fond of reducing his scientific reading to practice; and after studying Franklin's description of the lightning experiment, he proceeded to spend his store of Saturday pennies in purchasing about half a mile of copper wire. Having prepared his kite, he sent it up in the field opposite his father's door, and bringing the wire, insulated by means of a few feet of silk cord, over the backs of some of Farmer Wigham's cows, he soon had them skipping about the field in all directions, with their tails up. One day he had his kite flying at the cottage door as his father's galloway was hanging by the bridle to the paling, waiting for the master to mount. Bringing the end of the wire just over the pony's crupper, so smart an electric shock was given it that the brute was almost knocked down. At this juncture the father issued from the door, riding-whip in hand, and was witness to the scientific trick just played off upon his galloway "Ah! you mischievous scoundrel!" cried he to the boy, who ran off He inwardly chuckled with pride, nevertheless, at Robert's successful experiment.

After Franklin's audacious and most notable experiment, the history of the electric telegraph hastens apace. We have but to record one more experiment, before arriving at the date usually given as the first in the long series which indicates the development of telegraphs by electricity. This is 1787, when a Frenchman named Lomond succeeded in communicating signals from one house to another by electroscopic action.

It was in the year 1774 that George Louis Lesage, of Geneva, constructed a telegraph composed of twenty-four line wires, corresponding to the twenty-four letters of the alphabet, and by the use of frictional electricity and pith balls, succeeded in transmitting intelligible signals over the wires to a distance. The date mentioned is accordingly the time when the electric telegraph was invented, and Lesage was its inventor.

M. Lomond's name occurs once more among the eminent men to whom we are indebted for improvements previous to the introduction of the present system of rapid communication between widely different places, to which all precedent systems were but toys in comparison, although they were of great use in preparing the way for it.

The following passage occurs in "Arthur Young's Travels in France," published in Dublin in 1793. The date of the letter from which the extract is taken is Oct. 16th, 1787: "In the evening to Monsieur Lomond, a very ingenius and inventive mechanic, who has made an improvement in the jenny for spinning cotton. In electricity he has made a remarkable discovery. You write two or three words on paper; he takes it with him into a room, and turns a machine inclosed in a cylindrical case, at the top of which is an electrometer, and a small fine pitch ball; a wire connects with a cylinder and electrometer in a distant apartment, and his wife, by remarking the corresponding motions of the ball, writes down the words they indicate, from which it appears that he has formed an alphabet of motions. As the form of the wire makes no difference in the effect, the correspondence may be carried on to any distance within or without a fortified town, for instance, or for purposes much more worthy. Whatever the use may be, the invention is beautiful."

In the year 1794, M. Reiser, of Geneva, used thirty-six insulated wires for letters and numerals, in connection with a like number of narrow strips of tin foil pasted on glass; the letters and figures were cut in the foil and made visible by the passage of the electric spark. A year later, Tiberius Cavallo, in England, sent explosive and other electric signals through fine insulated copper wire, using Leyden jars, and sending "sparks at different intervals according to a settled plan." Three additional experiments, according to Steinheil, put it beyond a doubt that frictional electricity might be made a successful means of telegraphic intercourse. These were (1) That of D. F. Salva, Spain, who in the year 1798, worked an electric telegraph through twenty-six miles, using a single wire, and the sparks of a Leyden jar for signals. (2) That of Francis Ronalds, who, in 1816, constructed in England an experimental telegraph line, of a single insulated wire eight miles long, operated by an electrical machine, or small Leyden jar. His elementary signal was the divergence of the pith balls of a Canton's electrometer, produced by the communication of a statical charge to the wire. Lettered dials, rotated synchronously at each end of the line, served, in connection with the pith balls, to indicate the letter designated by the sender. This dial system was the precursor of Wheatstone's dial telegraph in 1839; House's letter printing telegraph in 1846; and Hughes' printing telegraph in 1855; and (3) that of Harrison Gray Dyar, America, who, in 1823, constructed a telegraph line on Long Island, supporting his wires by glass insulators fixed on trees and poles; the electric signals printed themselves upon litmus paper, the spacing of the marks indicating the letters and other signs. Just as Dyar and his partner Brown were seeking capital to set up a line between New York and Philadelphia, a blackmailing agent, failing to extort the concession of a large share in the enterprise, obtained a writ against the two partners on a charge of conspiracy to carry on secret communication between the cities! The case was never brought to trial, but the enterprise was blocked.

For the above information, beginning with the date 1794, we are largely indebted to an article which recently appeared in the *Scientific American*, reviewing a work on the origin and development of the electro-magnetic telegraph, with special reference to Professor Joseph Henry's contributions thereto. The work referred to is from the pen of William B. Taylor, an authority on the subject. We quote the remainder of the article as being the best summary of the subject with which we are acquainted, of particular use to the student of electricity, and of great value, for reference, to the general reader.

TELEGRAPHS BY GALVANISM.

"1808.—The first to apply to telegraphy the galvanic battery introduced by Volta, in 1800, was Dr. Samuel Thomas Von Soemmering, of Munich. He employed the energy of a powerful voltaic pile to bring about the decomposition of water by means of thirty-five gold pins immersed in an oblong glass trough. Each of these electrodes was in connection with one of the thirty-five wires forming the line. The bubbles evolved as these electrodes were received in figured and lettered tubes, and the messages were thus spelled out. In 1810 Soemmering telegraphed through two miles of wire.

"1816.—Dr. John Redman Coxe, of Philadelphia, suggested a system substantially the same as Soemmering's (of which he appeared to be ignorant). He also proposed to accomplish the same result by decomposing metallic salts, as was afterward done.

"1843.—Mr. Robert Smith, Scotland, devised a galvano-chemical telegraph, carrying out practically the suggestion of Dr. Coxe. At first he used a separate wire for each letter, the message being printed on a strip of paper wet with a solution of ferrocyanide of potassium. Subsequently Mr. Smith reduced his line to a single circuit of two wires, and worked his system through 1,800 yards of fence wire (1846).

"1846.—Mr. Alexander Bain, Scotland, patented in England a galvano-chemical telegraph, different in mechanical details, but similar in its chemical record to the system of Smith.

"1849.—Prof. Samuel F. B. Morse, New York, patented in this country a telegraph similar to Smith.

TELEGRAPHS BY GALVANO-MAGNETISM.

"1820.—Hans Christian Oersted, Copenhagen, rediscovered the directive influence of a galvanic conductor on a magnetic needle (Romagnosi's of the same in 1802 having attracted no attention). The same year (1820) Professor Schweiger, of Halle, made the first real galvanometer; and shortly afterward Ampere, in Paris, proved experimentally the feasibility of an electro-magnetic telegraph, in which the galvanometer should take the place of the electrometer employed by Lesage.

"1823.—Baron Paul L. Schilling, of Cronstadt, Russia, practically applied Ampere's suggestion. In his apparatus signals were produced by five galvanometer needles, provided with independent circuits.

"1824.—Peter Barlow, England, experimenting with considerable lengths of wire, to test the practicability of Ampere's suggestion, was convinced that it was impracticable, owing to the rapid dimunition of effect (due to increased resistance), by lengthening the conducting wire. Other inclusive experiments in the same direction were made by Fechter in 1829, and Ritchie in 1830.

"1833.—Prof. Carl Friedrich Gauss and Wilhelm Edward Weber constructed at Gottingen a galvanometer telegraph of a single circuit of uninsulated wire a mile and a half long. The alphabet of signs was made up of right and left deflections of the needle, observed by reflections from a small mirror. Gauss was the first to employ magneto electricity in telegraphs. Weber added to the signaling device a delicate apparatus for setting off a clock alarm.

"1836.—Prof. C. A. Steinheil, of Munich, undertook, at the request of Gauss, the development of the arrangement above described, and constructed a similar galvanometer telegraph line two miles in length, introducing considerable improvements. The next year Steinheil discovered that the ground might be made a part of the circuit, thus dispensing with a second wire for the return circuit.

"1837.—Mr. William Fothergill Cooke and Prof. Charles Wheatstone patented in England a galvanometer or needle telegraph very similar to the earlier one of Schilling, employing six wires and five indicating needles. An experimental line a mile and a quarter long was worked with partial success, July 25; and one thirteen miles long was established in 1838."

While these experiments with the needle were going on, the electro-magnet was being developed and applied.

1820.—The germ of the electro-magnet was discovered by Arago, who observed that the electric current would develop magnetic power in strips of iron and steel.

1824.—William Sturgeon, England, produced the true electro-magnet, with its intermittent control of an armature.

The electro-magnet of Sturgeon was improved by Professor Henry in 1828; and in 1829 he exhibited a larger magnet of the same character, tightly wound with 35 feet of silk covered wire. A pair of small galvanic plates, which could be dipped into a tumbler of diluted acid, was soldered to the ends of the wire, and the whole mounted on a stand. This was the first magnetic spool or bobbin. This invention was further improved the same year, and in 1830 Professor Henry, assisted by Dr. Philip Ten Eyck, constructed an electro-magnet which lifted 750 pounds. In 1831 he made one weighing $82\frac{1}{2}$ pounds, which sustained over a ton. In the meantime Professor Henry practically worked out the differing functions of quantity and intensity magnets, and experimentally established the conditions required for magnetizing iron at great distances through long conducting wires. This first made the electro-magnet available for telegraphic purposes.

1831.—The transmission of signals through a mile of copper bell wire interposed in a circuit between a small Cruickshank's battery and an intensity magnet—a practical telegraph—was practiced by Professor Henry.

This memorable experimental telegraphic arrangement involved three significant and important novelties. In the first place, it was the first electro-magnetic telegraph employing an "intensity" magnet capable of being excited at very great distances from a suitable "intensity" battery.

In the second place, it was the first electro-magnetic telegraph employing the armature as a signaling device, or employing the *attractive* power of the intermittent magnet, as distinguished from the directive action of the galvanic circuit. That is to say, it was, strictly speaking, the first magnetic telegraph.

In the third place, it was the first *acoustic* electro-magnetic telegraph.

1837.—Professor Samuel F. B. Morse devised a magneto-electric telegraph capable of transmitting signals through a circuit of forty feet, but failed for longer distances from the circumstance that he used a quantity current. His friend, Dr. Gale, made for him an intensity battery, and added a hundred or more turns to the coil of wire around the poles of the magnet. With these necessary (and radical) improvements the apparatus was made to work through ten miles of wire. In applying for a caveat for his invention, October 6, 1837, Professor Morse specified six distract parts, not one of which enters into the established "Morse" telegraph of to-day. Mr. Taylor shows that Professor

Morse's real contribution to telegraphy consists first in the adaptation of the armature of a Henry electro-magnet to the purpose of a recording instrument; and second, in connection therewith, the improvement on the Gauss and Steinheil dual-sign alphabets, made by employing the single line dot and dash alphabet.

In his general summary of the history of the origin and developement of the electro-magnetic telegraph, Mr. Taylor sets down the leading preparatory investigations and discoveries as these five :

1. The discovery of galvanic electricity by Galvani, 1786–1790.
2. The galvanic or voltaic battery by Volta, 1800.
3. The directive influence of the galvanic current on a magnetic needle by Romagnosi, 1802, and by Hoersted, 1820.
4. The galvanometer by Schweigger, 1820 (the parent of the needle system).
5. The electro-magnet by Arago and Sturgeon, 1820–1825 (the parent of the magnet system).

The second half dozen capital steps in the evolution of telegraphy were :

1. Henry's most vital discovery, in 1829 and 1830, of the intensity magnet and its intimate relation to the intensity battery.
2. Gauss' improvement, in 1833 (or probably Schilling's, considerably earlier), of reducing the electric conductors to a single circuit by the ingenious application of a dual sign, so combined as to produce

a true alphabet. (The anticipations of this idea by Lomond iñ 1787, Cavallo in 1795, and Dyar in 1825, are not regarded as practically influential in the progress of telegraphy).

3. Weber's discovery, in 1833, that the conducting wires of an electric telegraph could be carried through the air, without insulation, except at the points of support.

4. As a valuable adjunct to telegraphy, Daniell's invention of a constant galvanic battery in 1836.

5. Steinheil's discovery, in 1837, that a single conducting wire is sufficient for telegraphic purposes.

6. Morse's adaptation of the armature of a Henry electro-magnet as a recording instrument, 1837, and the single line dot and dash alphabet in 1838.

The earlier needle type of electro-magnetic telegraph has found its special application in ocean lines, no element of the Morse system entering into the operation of submarine cables.

The more recent telegraphic developments do not fall within the scope of Mr. Taylor's review. A few other dates, as given by Prescott, may appropriately serve to complete this chronology.

1861.—Reiss discovered that a vibrating diaphragm could be actuated by the voice so as to cause the pitch and rhythm of vocal sounds to be transmitted to a distance and reproduced by electro-magnetism.

1872.—Stearns perfected a duplex system, whereby two communications could be simultaneously transmitted over one wire.

1874.—Edison's quadruplex was invented.

1874.—Gray invented a method of electrical transmission, by means of which the intensity of tones as well as their pitch and rhythm could be reproduced at a distance; and subsequently conceived the idea of controlling the formation of electric waves by means of the vibrations of a diaphragm capable of responding to all the tones of the human voice.

1876.—Telephone invented.—Bell invented an improvement in the apparatus for the transmission and reproduction of articulate speech, in which magneto-electric currents were superposed upon a voltaic circuit, and actuated an iron diaphragm attached to a soft iron magnet. During the same year Dolbear conceived the idea of using permanent magnets in place of the electro-magnets and battery previously employed, and of using the same instrument for both sending and receiving.

1877.—Edison's carbon telephone was brought out.

To these may be added Edison's electro-motograph, or electro-chemical telephone, 1877.

1878.—Duplexing of ocean telegraph.

1879.—Cowper's writing telegraph.

1880.—Field's successful substitution of dynamo-electricity for galvanic batteries in telegraphing.

In the next chapter we shall introduce very interesting matter in regard to the early days of the electric telegraph, which, by the way, began in this country in 1844, with one wire between Baltimore and Washington.

INTRODUCTION OF THE ELECTRIC TELEGRAPH IN THE UNITED STATES.

Under this head we shall introduce matter personal, biographical and historical; funny and scientific—very miscellaneous, indeed, but all having a direct connection with that great event in our national history—the introduction of the electric telegraph into this country.

THE FIRST AMERICAN LINE.

Anent this event, it is a matter of historic record that on the 3d of March, 1843, Congress passed a bill appropriating thirty thousand dollars for the construction of Professor Morse's experimental line between Baltimore and Washington, in order to test the practicability of the invention. The original model of a telegraphic apparatus filed by the honored inventor when he got his patent has been unearthed from a lot of rubbish in the cellar of the Patent Office at Washington, where it has been lying for years. The clumsiness of the signal key, as compared with the one of the present day, is ridiculous. It is nearly two feet long, and has a large lump of lead at the furthest end from the hand, to throw the key and break the circuit. It was at first proposed to lay the wires under ground, inclosed in a leaden tube, and the contract for laying this tube was taken by Mr. F. O. J. Smith, of Maine, who was at that time editor of the Portland *Farmer,* and who had previously been—as a member of Congress, and chairman of the Committee on Commerce—largely instrumental in the passage of the appropriation. About this time Mr. Ezra Cornell, who was on a visit to Maine on business, called upon Mr. Smith, who, in speaking of the contract which he had taken for laying the wires, and for which he was to receive one hundred dollars per mile, incidentally remarked that, after a careful examination, he had found he would lose money by the job. Mr. Smith at the same time showed Mr. Cornell a piece of the pipe, and explained the manner of its construction, the depth to which it was to be laid, and the difficulties which he expected to encounter in carrying out the design. Mr. Cornell at this same interview, after the brief explanation which Mr. Smith had given, told him that in his opinion the pipe could be laid by machinery at a much less expense than one hundred dollars per mile, and would be in the main a profitable operation. At the same time he sketched on paper the plan of a machine which he thought practicable. This led to the engagement of Mr Cornell by Mr. Smith to make such a machine, and he immediately went to work and made patterns for its construction. While the machine was being made, Mr. Cornell went to Augusta, Maine, and settled up his business, and then returned to Portland and completed the pipe machine. Professor Morse was notified by Smith in regard to the machine, and went to Portland to see it tried. The trial proved a success. Mr. Cornell was employed to take charge of laying the pipe. Under his hands the work advanced rapidly, and he had laid ten miles or more of the pipe when

Professor Morse discovered that the insulation was so imperfect that the telegraph would not operate. He did not, however, stop the work until he had received orders; which order came in the following singular manner: When the evening train came out from Baltimore, Professor Morse was observed to step from the car; he walked up to Mr. Cornell, took him aside, and said: "Mr. Cornell, cannot you contrive to stop the work for a few days without its being known that it is done on purpose? If it is known that I ordered its stoppage, the plaguy papers will find it out and have all kinds of stories about it." Mr. Cornell, with his usual quickness of discernment, saw the condition of affairs and told the professor that he would make it all right. So he ordered the drivers to start the team of eight mules which set the machine in motion, and, while driving along at a lively pace, in order to reach the Relay House, a distance of about twenty rods, before it was time to "turn out," managed to tilt the machine so as to catch it under the point of a projecting rock. This apparent accident so damaged the machine as to render it useless. The professor retired in a state of perfect contentment, and the Baltimore papers on the following morning had an interesting subject for a paragraph. The work thus being of necessity suspended, Professor Morse convened a grand council at the Relay House, composed of himself, Professor Gale, Dr. Fisher, Mr. Vail, and F. O. J. Smith, the persons especially concerned in the undertaking. After discussing the matter, they determined upon further efforts for perfecting the insulation. These failed, and orders were given to remove everything to Washington. Up to this time Professor Morse and his assistants had expended twenty-two thousand dollars, and all in vain. Measures were taken to reduce the expenses, and Mr. Cornell was appointed assistant **superintendent**, and took entire charge of the undertaking. He now altered the design, substituting poles for the pipe. This may be regarded as the commencement of "air lines" of telegraph. He commenced the erection of the line between Baltimore and Washington on poles, and had it in successful operation in time to report the proceedings of the conventions which nominated Henry Clay and James K. Polk for the presidency.

APATHY OF SCIENTISTS, PRESS AND PUBLIC.

Although the practicability of the telegraph had been so thoroughly tested, it did not at once become popular. A short line was erected in New York city in the spring of 1845, having its lower office at 112 Broadway and its upper office near Niblo's. The resources of the company had been entirely exhausted, so that they were unable to pay Mr. Cornell for his services, and he was directed to charge visitors twenty five cents for admission, so as to raise the funds requisite to defray expenses. Yet sufficient interest was not shown by the community even to support Mr. Cornell and his assistant. Even the New York press was opposed to the telegraphic project. The proprietor of the New York *Herald*—think of the astute elder Bennett making such a big blunder—when called upon by Mr. Cornell and requested to say a good word in his favor, emphatically refused, stating distinctly that it would be greatly to his disadvantage should the telegraph succeed. Stranger still it is that many of those very men who would be expected to be entirely in favor of the undertaking, namely, men of scientific pursuits, stood aloof and declined to endorse it. In order to put up the line in the most economical manner, Mr. Cornell desired to attach the wires to the city buildings which lined its course. Many house-owners objected, alleging that it would invalidate their insurance policies by increasing the risk of their buildings being struck by lightning. Mr. Cornell cited the theory of the lightning rod as demonstrated by Franklin, and showed that the telegraphic wire would add safety to their buildings. Some persons still refused, but informed him that could he procure a certificate from Professor Renwick, then connected with Columbia College, to the effect that the wires would not increase the risk of their buildings, they would allow him to attach his wires. Mr. Cornell thought the obtaining of such a certificate a very easy matter, and certainly all scientific men were agreed upon the Franklin theory. He therefore posted off to Columbia College, saw the distinguished savan, stated his errand, and requested the certificate, saying it would be doing Professor Morse a great favor. To his utter consternation the learned professor replied: "No, I cannot do that," alleging that "the wires *would* increase the risk of the buildings being struck by lightning." Mr. Cornell was obliged to go into an elaborate discussion of the Franklin theory of the lightning-rod, until the professor confessed himself in error, and prepared the desired certificate, for which opinion he charged him twenty-five dollars. This certificate enabled Mr. Cornell to carry out his plans.

DESCRIPTION OF THE INSTRUMENTS FIRST USED.

The apparatus used on the original line between Baltimore and Washington in 1844 would be something of a curiosity at the present time. The relay magnets weighed one hundred and eighty-five pounds, and it required two men to handle one of them if it became necessary to move it. The coils were about eighteen inches in diameter, and were composed of No. 16 copper wire insulated with cotton thread. It was supposed at that time to be indispensably necessary that the wire surrounding the magnets should be the same size as the wire of the line. Professor Charles Grafton Page, a short time afterward, devised a magnet of considerably less size, which was used in the lines built during the years 1845 and 1846. Professor Morse, while in France in the year 1845, obtained some electro-magnets of about the same size of those now in use, which he brought to this country and made use of in working the telegraph. The first small relay magnet made in this country was constructed, we believe, by Clark of Philadelphia in 1845 or 1846, and in its general form was very similar to those now in use.

An interesting relic of the early days of telegraphy has been discovered at Morristown, N. J. It is the first instrument by which messages were received and sent by aid of the electric current, and was one of two taken from Morristown by Morse and Vail—Morse using one at Washington, and Vail the other at Baltimore. The first message sent was the now well-known "What has God wrought?" which Morse transmitted to Vail; but the first public message was the news of the nomination of Polk to the presidency by the Baltimore convention of 1844, sent by Vail to Morse.

These instruments were in constant use for six years, when Mr. Vail, returning to Morristown, brought his with him, and where it has still remained in the possession of his family. Mr. Vail dying soon after, his instrument was specially left by a clause in his will to his eldest son as an heirloom, while parts of instruments made during the experimental trials were left to Professor Morse, with a request that he would give them at some future day to the New Jersey Historical Society. The old instrument works as well as when first made. Its dimensions are sixteen inches in length, seven inches in height, six inches wide, with two magnets of three inches diameter. The paper used was two and a half inches in width, three pens being proposed to be used. The weight of the instrument is twenty pounds.

"GREAT OAKS FROM LITTLE ACORNS GROW."

In the year 1850, Mr. Alfred Vail, of whom we shall have occasion to say more by and by, wrote a manuscript giving an account of the receipts of the telegraph at the Washington office during the first four days of its operation after it had been taken under the patronage of the government. The details form a forcible illustration of the truth of the motto at the head of this paragraph. Mr. Vail's manuscript reads as follows:

MR. VAIL'S ACCOUNT OF THE FIRST WEEK OF THE TELEGRAPH.

"The telegraph was first put in operation between Washington and Baltimore in the spring of 1844, and was shown without charge until April 1, 1845. Congress, during the session of 1844-45, made an appropriation of $8,000 to keep it in operation during the year, placing it, at the same time, under the supervision of the postmaster-general. He, at the close of the session, ordered a tariff of charges of one cent for every four characters made by or through the telegraph, appointing also the operators of the line—Mr Vail for the Washington station, and Mr. H. J. Roberts for Baltimore.

"This new order of things commenced on April 1, 1845, and the object was to test the profitableness of the enterprise. The receipts for April 1-4, inclusive, were as follows:

"It should be borne in mind that Mr. Polk had just been inaugurated, and, as is always the case on the advent of a new administration, the city was filled with persons seeking for office. A gentleman of Virginia, who stated that to be his errand to the city, came to the office of the telegraph on the 1st day of April, and desired to see its operation. The oath of office being fresh in the mind of the operator, and he being determined to fulfil it to the letter, the gentleman was told of the rates of charges, and that he could see its operation by sending his name to Baltimore and having it sent back, at the rate of four letters or figures for a cent, or he might ask Baltimore regarding the weather, etc. This he refused to do, and coaxed, argued and threatened. He said there could be no harm in showing him its operation, as that was all he wanted. He was told of the oath just taken by the incumbent, and of his intention to keep it faithfully; and that, if it was shown to him by the passage of a communication gratuitously, it would be in violation of his oath of office. He stated he had no change. In reply, he was told that if he would call upon the postmaster-general and obtain his consent that the operation should be shown him gratis, the operator would cheerfully comply to almost any extent. He stated in reply that he knew the postmaster-general, and had considerable influence with some of the officers of the government, and that he (the operator) had better show it to him at once, intimating that he might be subjected to some peril by refusing. He was told that no regard would be paid to the extent of his influence, be it great or little; that he did not think he was at liberty to use the property of the government for individual benefit when under oath to exact pay, and cited the rules of the post-office in relation to the carriage of letters, but that he was willing to do as directed by the postmaster-general (Hon. Cave Johnson). The discussion lasted almost an hour, when the gentleman left the office in no pleasant mood.

"This was the patronage received by the Washington office on the 1st, 2nd, and 3rd of April. On the 4th the same gentleman turned up again, and repeated some of his former arguments. He was asked if he had seen the postmaster-general, and obtained his consent to his request, to which he replied he had not. After considerable discussion, which was rather amusing than vexatious, he said that he had nothing less than a twenty-dollar bill and one cent, all of which he pulled out of his breeches pocket. He was told that he could have a cent's worth of telegraphing, if that would answer, to which he agreed. After his many manœuvres and long agony the gentleman was finally gratified in the following manner: Washington asked Baltimore 4. which meant, in the list of signals, 'What time is it?' Baltimore replied 1, which meant '1 o clock.' The amount of the operation was one character each way, making two in all, which, at the rate of four for a cent, would amount to half a cent exactly. He laid down his cent, but was told that half a cent would suffice, if he could produce the change. This he declined to do, and gave the whole cent, after which, being satisfied, he left the office.

"Such was the income of the Washington office for the first four days of April, 1845. On the 5th twelve and a half cents were received. The 6th was the Sabbath. On the 7th the receipts ran up to sixty cents; on the 8th to $1.32; on the 9th to $1.04. It is worthy of remark," concludes Mr. Vail, "that more business was done by the merchants after the tariff was laid than when the service was gratuitous."

The humors of the telegraph form a fruitful subject. Numerous good stories are constantly cropping out to vary and relieve the routine of telegraphic operations. Many of these get into the public prints, and increase that capital of mirthful yarns which is an important means of health to this over-worked generation. Upon comparing the best of these with those that are told of the funny blunders and incidents accompanying the beginning of telegraphic operations, the writer is of the opinion that the latter are certainly the more side splitting Take for instance that of the old lady who wrote a letter to headquarters asking them to remove the wires which had been attached to her chimney, and said: "I must request you to remove your wires from my chimney immediately. The noise the message makes going along the wires is sometimes awful, and sometimes—I suppose when the operator has a hard word to spell—I declare it quite shakes the house."

Another good old soul, with due respect for the proprieties, on seeing some telegraph wire while taking her first ride by rail, was heard to remark: "Well, I have often said they would never git me into the railroad cars, but I *know* they will never git me on to them telegraft wires."

When the telegraph was being introduced into a populous district of Massachusetts, hardly five minutes had elapsed after the erection of one of the poles, before some enterprising genius posted a bill thereon, and soon two street Arabs were attracted to the spot, when the following dialogue ensued:

"I say, Mickey, what an invintion the telegraph is."

"Yes, an' here's a dispatch broke out on the post a'ready."

When the telegraph was first put in operation between Portland and Boston, a countryman drove a flock of turkeys to the former place for a market, but not finding so good a sale as he anticipated, he inquired of some by-standers their price in Boston. Some wag of a fellow advised him to step into the telegraph office. Jonathan entered and put the all-important question to the operator, who immediately telegraphed to Boston, and in a few minutes received an answer to his inquiry, and informed his customer. Jonathan looked at the operator with a sly wink and exclaimed: "You can't gum it over me." He was about leaving the office when the operator told him that there were nine shillings to pay. Jonathan bristled up and burst forth in a rage: "You can't gum it over me. That old tick-box of yourn hain't been out of this room since I've been here."

Notwithstanding the severest kind of temptation, the humor reserved for this place must be cut down to one more story, told at the expense of Mr. J. B. Stearns, who afterward invented the Stearns system of duplex telegraphing, referred to in last chapter. Mr. Stearns at that time officiated as one of the operators in the now long ago when the Boston fire alarm was a new institution. One summer night when he was on duty, he was startled by hearing a church bell in South Boston, which was connected with one of the "alarm circuits," break forth at a most unseemly hour with a continuous "ding dong, ding-dong," which bade fair to awaken every slumbering inhabitant within the radius of a mile. As the striker of the bell was driven by water power, which was merely controlled by the electric current, Stearns was fully aware of the fact that the armature of the magnet had "stuck," or otherwise got out of gear, and that the racket would probably continue until some one went over there and adjusted it, or else until the supply of Cochituate water failed—a slightly improbable contingency. Of course he couldn't leave his post, and therefore was obliged to sit and listen to the concert, which, under the circumstances, he probably enjoyed nearly as well as the citizens in the immediate neighborhood of the performance. Stearns, however, being a gentleman of resources, was not to be foiled so easily. A happy thought finally struck him. He would reverse the battery on that circuit, which would doubtless release the "stuck" armature, and restore quiet to the distracted inhabitants, who by this time were doubtless beginning to get mad, and revile the fire alarm and the individuals connected therewith in a highly improper manner. The wires were changed, and the clamor instantaneously ceased. On the following morning, in the serene consciousness of a good deed well performed, Stearns duly reported the incident to his chief, Mr. Moses G. Farmer, who did not hesitate to bestow the praise so justly due to the ingenuity of his subordinate, but suggested that it might also be well to examine the defective striker, and, if need be, adjust it, so as to prevent the possibility of another similar accident. Singular to state, when Stearns arrived at the scene of the previous night's disturbance he found that the door of the church, and also the belfry, had been stove in with an axe, and the hammer of the bell effectually tied up by a strong rope. Whether this result was, as a whole, brought about merely by the reversal of that battery still remains an open question. It is understood, however, that even Stearns himself has always entertained some doubt of it.

"HONOR TO WHOM HONOR IS DUE."

Clamors are still made by the respective friends of those who aided in bringing the telegraph to perfection, for the preference to be given to a certain one out of several names conspicuous among the leaders in, possibly, this highest achievement of human ingenuity.

RONALDS.

England has lately witnessed the conferring of the honor of knighthood upon Mr. Francis Ronalds, for whom no meaner an authority than the *Pall Mall Gazette* claims that he "is neither more nor less than the originator of our telegraph system. He was the very first," it adds, "either in England or abroad, to invent an electric telegraph so constructed as to be capable of extensive practical application, and so far back as 1823 he fully developed its principle and mode of action. Still earlier, namely, in 1816, he had constructed a working electric telegraph, and on offering it to the then government, received an answer which can never be too often cited as an illustration of official complacency:

'Telegraphs of any kind are now wholly unnecessary, and no other than the one now in use will be adopted.' Nothing daunted by this apathy, Mr. Ronalds matured his invention, and in 1823 published a 'Description of an Electric Telegraph, and of some other Electrical Apparatus.' Mr. Ronalds was too far ahead of his time, and too purely a man of science, to secure a hearing for his discovery in those early days, and it was left to others to mature his idea, and to establish the system which his prophetic eye had foreseen would one day transform the world. It was not till 1837, fourteen years after Mr. Ronalds' pamphlet, that Messrs. Cooke and Wheatstone took out their first patent. The science and practical skill of these and other eminent electricians have brought electric communication to its present state; but the great fact remains that Mr. Ronalds was the first to demonstrate practically the principle which they have developed."

MORSE.

In our own country, Morse has been, as is usual in such disputes, both unduly praised and undervalued on account of his services to the public.

Readers of the foregoing matter, particularly the concluding portion of an earlier chapter and the beginning of this, are able, it is believed, to form a clear conception of what he did, stated as matters of fact, without exaggeration or depreciation. In the opinion of the writer, the public honors paid Morse, notably in the erection of his statue in Central Park, have in no wise exceeded his merits; but there may be room, he thinks, for the more emphatic recognition of services rendered by gentlemen associated with him in the introduction of the same wonderful improvement.

HENRY.

For the purpose of doing justice to a name which cannot be held in too great honor, it is but right to append to these observations a summarized statement of what Professor Henry did toward the development of the telegraph. His improvement upon Sturgeon's electro-magnet "consisted in insulating the conducting wire itself, instead of the rod to be magnetized, and covering the whole surface of the iron with a series of coils, in close contact. Henry's magnet was described in Silliman's *Journal* in 1831; and, in 1832, a mechanical arrangement was put up in the Albany Academy for making signals and sounding a bell through a wire more than a mile in length. Previous to Professor Henry's investigations the means of developing magnetism in soft iron were imperfectly understood, and no electro-magnet, applicable to the telegraph, was known The particular form of battery adapted to project the current through a long conductor was first pointed out by Henry, and he was the first to magnetize a piece of iron at a distance, and to call attention to the fact of the applicability of the experiment to the telegraph The principles developed by him were applied to render the various machines invented by Gauss, Weber, Steinheil, Wheatstone and Morse effective at a distance. The galvanometer now employed for transmitting messages by the Atlantic cable, is about as close an imitation of the apparatus devised by Henry for ringing a bell, in the Albany Academy in 1832, as the different circumstances of the cases require; and the electro-magnet, now used for the telegraph all over the world, is the one invented and described by Henry in 1831. Whether the instrument used be a semaphore—that is, carrying evanescent signals, or a telegraph making a permanent record—the engine for driving the works by aid of the battery is the electro-magnet invented by Professor Henry The philosopher who discovered the scientific principles upon which the electro-magnet is founded, and who invented the form of apparatus best adapted to demonstrate these principles, must be regarded by the whole world as having made the chief contribution toward the application of electro-magnetism to the various wants of man This philosopher was Joseph Henry. and to him was accorded the homage of the whole scientific world for his magnificent researches." So speaks the *Scientific American* in an article whose principal points we hereby gladly assist to preserve.

VAIL.

Not long ago Cincinnati brought to light a volume of nearly eight hundred pages, entitled· "Up the Heights of Fame and Fortune, and the Routes Taken by the Climbers to become Men of Mark," containing, among much interesting matter, notices of Professor Morse and his associate, Alfred Vail. To the latter is ascribed the invention of the first available telegraph instrument. Mr Vail was born in New Jersey in 1807, and early displayed great mechanical ingenuity. While a student under Prof. Morse at the New York University, in 1835, he saw the latter's first rude machine, and, by virtue of engaging to devote his personal services and skill toward perfecting the invention, became an owner of one-eighth of the patent. He also offered Prof. Morse much needed pecuniary aid. In 1853 the professor said that to the joint liberality of Vail's father and brother, "but especially to Alfred's attention and skill and faith," was due the success of his early endeavors to bring the telegraph before the public. On the passage of the telegraph bill in 1843 Alfred was appointed one of Mr. Morse's assistants.

Having stated these biographical facts, the writer of the book we have referred to, says:

"The Morse machine of 1836 passed into Vail's hands in 1837, for an entire mechanical reconstruction throughout—to speak a language not entirely unknown to the first machine, but to perform entirely new functions, and to produce an entirely new system of signs and letters which the first by its structure was physically unable of being made to speak. Alfred Vail invented the first combination of the horizontal lever motion to actuate a pen, pencil or style, and the entirely new telegraphic alphabet of dots, spaces and marks, and he did so prior to September, 1837, the month when the old instrument passed into his hands for reconstruction. His more perfect invention of a steel style upon a lever, which could strike into the paper as it was drawn onward over a ground roller, and emboss upon it the same alphabetic characters, was not invented until 1844, about the time the first line of telegraph began to operate between Baltimore and Washington. This instrument, somewhat transformed, still holds its place as practically the best ever invented."

He then quotes extensively from the correspondence between Professors Morse and Henry, and reproduces a plain-spoken letter from Vail, to show why he did not urge his claims to the credit of the invention. This letter is deemed of so much importance by the publishers that they have given it in lithographic *fac-simile*. It is as follows:

"The lever and roller were invented by me, in the sixth story of the New York *Observer* office, in 1844 before we put up the telegraph line between Washington and Baltimore, and this *combination* has been always used in Morse's instrument. I am the sole and only inventor of this mode of telegraph embossed writing. Professor Morse gave me no clue to it, nor did any one else, and I have not asserted publicly my right as first and sole inventor because I wished to preserve the peaceful unity of the invention, and because I could not, according to my contract with Professor Morse, have got a patent for it.

"ALFRED VAIL."

As early as 1847 Prof. Morse urged Mr. Vail to sell to him his interest in the telegraph for $15,000, but he refused. He died in January, 1859. Amos Kendall, a friend of both parties, said: "If justice be done, the name of Alfred Vail will forever stand associated with that of Samuel F. B. Morse, in the history of the invention and introduction into public use of the electro-magnetic telegraph."

A CLASS WHOSE SERVICES SHOULD NOT BE OVERLOOKED.

In giving "honor to whom honor is due," we must not overlook the claim of the laborer for the recognition of his indispensable services in making the telegraph a public convenience. A young Irishman, a member of a debating society in Geneva, New York, may be quoted as having done good service to his class by his emphatic and characteristically "bulling" method of making his claim. At a recent meeting of his society the subject of discussion was: "Which is of the most benefit to the country—the mechanic or the laborer?" One young man took the side of the mechanic, and expatiated at great length. Among a multitude of other things, he claimed that mechanics made and laid the Atlantic cable, and sat down amid loud applause. For a few minutes it looked as if there was no one bold enough to challenge his conclusion. At length a laborer came forward and said that he had a few words to say on the subject. He was willing to admit that the mechanic had made and laid the Atlantic cable; but, exclaimed he, smiting the table with a fist about the size of a twenty-three pound ham, and looking around with an air of triumph upon the audience, who were terrified at seeing the table sink to the floor under the force of his ponderous blow: "Be jabers, who dug the post holes?"

A CHAPTER ABOUT OPERATORS AND MES-SENGERS.

The electric telegraph has created a new industry, in its nature pleasantly intermingling manual and mental operations, not severe, but requiring close attention; educational of the observation and judgment, and affording scope to the ambitious for remunerative promotion. Moreover, it gives employment to women as well as men, and thus assists in the practical solution of the difficult question: What must society do with the capable and intelligent female population who cannot marry, for the very sufficient reason, among others, that there are not enough men to mate every one of them? The army of bright boys employed as messengers must not be lost sight of here. These earn their living in a manner which gives them enough physical and not too great mental exercise; as desirable as any, in short, for quick and growing boys, many of whom themselves subsequently become professional operators, or, if not, are at least prepared, by their apprenticeship in the telegraph office, for other useful employment.

THE TELEGRAPH OFFICE A SCHOOL FOR THE STUDY OF HUMAN NATURE.

While, in common with all other occupations, that of the telegraph operator is one of detail and routine, it probably affords more than any other, in the variety of people requiring his assistance, and the diverse character of messages received and sent, matter of amusement, information, thought and reflection. The telegraph office is a school for the study of human nature, of the multifarious occasions of business, the domestic and social relations, and politics, and the mental and emotional operations called out thereby. In the words of a writer whose name we regret to be unable to give: "The telegrapher's window is an eye through which the operator looks upon the world. Before it passes in a single day more of the very wine of human experience than one could observe in a whole decade of European travel. The business man, brisk, keen and active, leers at him through that window; the burglar, bold and skillful, sends his telegram in cipher to a confederate; and the widow, in weeds, sends to her friends the mournful sentences: 'Charley is dead. Come to me!' The telegrapher receives the communication respectfully, duly marks it with some hieroglyphic signs, and speedily the electric soul of the battery utters, a thousand miles away: 'Charley is dead. Come to me!' It may be to a mother, to a father, or to a brother; but it carries a pressing request, and to-morrow, or the day after, the individual to whom the message is addressed is in New York. Or it may be that the father, or mother, or sister, or brother, cannot leave home; and then comes back the sorrowful answer: 'Business is pressing; will come as soon as I can.' And the widow weeps alone with her dead.

"Curious messages in curious handwriting are handed to him through the window—telegrams with bad spelling, and telegrams with bad grammar; telegrams that a hieroglyphicist, who may have delved for years amid the mummy-cases of Egypt, could never unriddle; and these last are handed back with a suave request to read and interpret.

"There are telegrams in cramped, unnatural hand, and telegrams in the round, fanciful hand of the writing-master; telegrams with capitals where they should not be, and telegrams with no caps' at all—but very few with 'caps' where they should be; telegrams of laborious pomposity from venerable professors, and telegrams curt and brief and epigrammatic, from those who know how to save a penny at the expense of perspicuity; in short, there are telegrams of all sorts—not excepting dead-head telegrams, of which some are sent and some are not sent, according to the claims of the individual to be considered a dead-head."

CONSCIENTIOUS CARE GIVEN TO HAVING MESSAGES DELIVERED.

Operators, as a class of public servants, are among the worthiest. Outsiders know little or nothing of the pains sometimes taken, without request or remuneration, to insure the delivery of dispatches incorrectly addressed. This is mentioned only as an example of the conscientious care given to work which comes before them, in many cases where in strict justice, it might be laid aside without prejudice to the operator's interests with relation to his employer, as being in accordance with the prescribed routine of the office.

THE LITERATURE OF THE TELEGRAPH.

The operators of the United States are justly proud of

their professional skill, and generously assist the means of literary communication afforded them in the books and papers prepared for their use and recreation. These evidence a very considerable degree of literary merit, and there is an intelligent demand for mo'e which reflects credit upon the craft. The literature of the telegraph is a most interesting and pleasing feature of the times.

Curious examples have been given of operators communicating by Morse characters under circumstances of peculiar difficulty. The most notable of these which the writer ever encountered is the following, from the pen of Mr. D. B. Grandy, until recently a well-known operator in the Boston office. It is proper to say that, with a view of verifying the statement before giving it here, the publisher wrote Mr. Grandy on the subject, and received the reply that the matter was precisely as given in the subjoined account:

A CASE OF EPILEPTIC TELEGRAPHY.

"In the winter of 1870-71," says Mr. Grandy, "I was employed in the Western Union office at Boston. Among my associates was George ———, with whom I had formed an intimate acquaintance and friendship. One evening I was at the theatre, when considerable commotion occurred in the balcony above me. After the play I learned that a man had fainted and been carried out insensible. On arriving at my lodgings I found that the man was no other than my friend George, who also occupied a room in the same house. I went to his room and found his room-mate and a physician there, while George lay on the bed, his face pale, his eyes open, but fixed and glassy, and his limbs cold and rigid as death. The physician pronounced it an epileptic fit. We spoke to him, chafed him, and made every effort to rouse him, but in vain. Finally we sat down and awaited his return to consciousness. I drew my chair up to his side, and took his hand in mine. As I did so I noticed a feeble pressure by his fingers, and then that pressure resolved itself into dots and dashes, and I read from them:

"'W-h-a-t d-o-c-t-o-r s-a-y a-b-t m-e?'

"'I asked him if he could hear what I said to him.'

"'Y-e-s.'

"'Are you in pain?'

"'Y-e-s.'

"'Can't you speak?'

"'N-o.'

"In short, I got, from the slight pressure of his fingers, enough dots and dashes to describe his feelings to the physician, who was enabled by the description thus obtained to judge of his condition and apply the necessary remedies, so that, after watching by his bedside until the small hours of the morning, we were relieved from our anxiety by signs of returning animation. By four o'clock he was completely himself again, but greatly exhausted, and it was several days before he was able to appear at the office. He afterward informed me that from the time he fainted in the theatre until he came out of the trance, he knew all that was passing around him, and heard all that was said, but could neither see, speak nor move a muscle, except those of his fingers, which he was able to use sufficiently to communicate with us by feeble dots and dashes. The physician pronounced it the most singular case of the kind that ever came under his treatment. Certainly no other method of communicating was possible in his condition, and it would seem from this incident that a person in a dying condition would be able, if he possessed a knowledge of telegraphic characters, to let his thoughts and feelings be known long after any other means of communication became impossible."

AN ARMLESS OPERATOR.

Mr. Patrick Shea, of Binghamton, N. Y., operates without arms, an accomplishment mastered after six months of close and unwearied application. Having lost both arms in an accident while fireman on the Albany and Susquehanna Railroad, he was provided with a pair of cork substitutes, and with these performs all his duties as operator.

A DEAF OPERATOR RECEIVING BY SOUND.

When the magnetic telegraph was first introduced, there was an arrangement by which the letters and words communicated were reeled off by means of punctures in long narrow strips of white paper, after this fashion, namely : - —, — - -, etc. These were translated by the receiving operator, and thus rendered into readable English. In the course of time this attachment to the Morse instrument was dispensed with, and the operators, instead, read the messages by sound, or the clicking of the instrument, with the proper intervals for a clear understanding of that language, so that there could be just as much certainty as there is in speaking, compared to written or printed communications. The operator's ear was rendered more and more acute, and he, therefore, could hear the faintest vibrations, or the whisperings of the instrument. But one would scarcely think that the arrangement would suit a deaf man. It does not, but the deaf man can suit himself to even these circumstances. The fact is demonstrated. There was a gentleman in the American Telegraph Company's office, in Washington, who, though he could not hear, was classed as a first-class operator, dealing with sounds! He could send and receive dispatches intelligently. But how was this done? By the sense of *feeling*. He placed his leg against that of the instrument table, and in other ways read by the slight jarring, while watching the operation of the instrument itself, and he thus understood all that the little "sounder" was talking about.

A "FRISCO" YARN.

"Two young men," says the *Chronicle* of San Francisco, "telegraph operators, board at one of our leading third-class hotels, and being of a somewhat hilarious disposition, find great amusement in carrying on conversation with each other at the table by ticking on their plates with a knife, fork or spoon. For the information of those not familiar with telegraphy it may be well to state that a combination of sounds or ticks constitutes the telegraphic alphabet, and persons familiar with these sounds can converse thereby as intelligibly as with spoken words. The young lightning strikers, as already stated, were in the habit of indulging in table talk by this means whenever they desired to say anything private to each other. For instance, No. 1 would pick up his knife and tick off some such remark as this to No. 2: 'Why is this butter like the offence of Hamlet's uncle?'

"No. 2—'I give it up.'

"No. 1—'Because it's rank, and smells to heaven.'

"Of course the joke is not appreciated by the landlord (who sits close by), because he doesn't understand telegraphic ticks, and probably he wouldn't appreciate it much if he did; but the jokers enjoy it immensely, and laugh immoderately, while the other guests wonder what can be the occasion for this merriment, and naturally conclude that the operators must be idiots.

"A few days ago, while these fun-loving youths were seated at breakfast, a stout-built young man entered the dining room with a handsome girl on his arm, whose timid, blushing countenance showed her to be a bride. The couple had, in fact, been married but a day or two previous, and had come to San Francisco from their home in Oakland, or Mud Springs, or some other rural village, for the purpose of spending the honeymoon. The telegraphic tickers commenced as soon as the husband and wife had seated themselves.

"No. 1 opened the discourse as follows: 'What a lovely little pigeon this is alongside of me—ain't she?'

"No. 2—'Perfectly charming—looks as if butter wouldn't melt in her mouth. Just married, I guess. Don't you think so?'

"No. 1—'Yes, I should judge she was. What luscious

lips she's got! If that country bumpkin beside her was out of the road, I'd give her a hug and a kiss, just for luck.'

"No. 2—'Suppose you try it anyhow. Give her a little nudge under the table with your knee.'

"There is no telling to what extent the impudent rascals might have gone but for an amazing and entirely unforseen event. The bridegroom's face had flushed, and a dark scowl was on his brow during the progress of the ticking conversation, but the operators were too much occupied with each other to pay any attention to him. The reader may form some idea of the young men's consternation when the partner of the lady picked up *his* knife and ticked off the following terse but vigorous message:

"'This lady is my wife, and as soon as she gets through with her breakfast I propose to wring your necks, you insolent whelps.'

"The countenances of the operators fell very suddenly when this message commenced. By the time it ended they had lost all appetite and appreciation of jokes, and slipped out of the dining room in a very rapid and unceremonious manner. The bridegroom, it seems, was a telegraph operator himself."

RECOGNIZING EACH OTHER'S TOUCH.

Operators who are in the habit of receiving from and sending to one another, become so accustomed to the peculiarities of each other's touch as to readily recognize it. For example, it is told of Mr. Hempstead, one of the operators in the Western Union Telegraph Company's office at Hartford, Conn., that by this means he succeeded in making a discovery of great importance to an unfortunate man and his friends. The circumstances were these: Mike W. Sherman, formerly a telegraph operator in Hartford, escaped from the Middletown insane asylum, where he had been confined, and, though thorough search was made for him, he for about two weeks successfully eluded those who were on his track. While Hempstead was at work in the Hartford office one night he suddenly recognized, among the clatter of a score of messages passing over the wire, a sound which he at once declared was the touch of the missing Mike. It proved to be a message from Wallingford, and an investigation showed that the Hartford operator was quite right in ascribing it to the insane man, who was afterward found there, he having dropped into the office in the former place, and taken a hand at his old business.

This same ability of distinguishing touch is a means of friendly intercourse between operators separated by long distances, and who probably have never seen each other. Attracted by an influence more subtle than the electric fluid itself, lovers have formed their first intimacy by this means, and not always with the ill-fortune which it appears followed the " Misplaced (Telegraphic) Affection "—shall we say "immortalized," by Beta, in a rhyming effusion which first saw the light as a contribution to *The Telegrapher.*

MISPLACED (TELEGRAPHIC) AFFECTION.

Thomas Tot, telegraphist, ten hours every day
Labored conscientiously for promises to pay;
On the self same circuit, not a thousand miles from T,
Nancy Anna Wilkins gently jerked the mystic key.

What could be expected when we note their common labors?
What, when we consider that the two had long been neighbors?
(Not so near that they had met, but near enough, 'tis true
Little distances may lend enchantment to a view.)

What could be expected under all the circumstances,
But that each should halo each with tender loving fancies?
But that each in painting each should color each in glory?
What could be anticipated—save the old, old story?

She, in his imaginings, lived something light and airy,
Like "Sweet Home," or cotton wool, a zephyr or a fairy;
He, in hers, existed something big, bold, loud, defiant,
Brave as Jack the Killer and as burly as the giant.

Nancy fell in love with Thomas Tot's manipulation;
He could take and shake a key to whip the 'tarnal nation;
He could send—you all must know what merit there was in it—
Eighty, more or less, and "take" some ninety words per minute.

Thomas fell in love with Nancy Anna's disposition,
You yourselves had done the same if placed in his position;
O, she was—by telegraph—as sweet as Jersey peaches,
With a knack for simple jokes and sentimental speeches.

Every week-day morning, when the wires were in trim,
Thomas said $g\ m$* to her, and she $g\ m$ to him;
Every idle afternoon when business was over,
Down they sat to have a chat, and thought themselves in clover.

Many years of this rolled on in regular rotation,
'Till came round Tom's decreed two weeks' (or less) vacation;
So he telegraphed his friend to don her silks and satins,
For that he would be with her before the morrow's matins.

Nancy Anna decked herself in everything that glitters,
Fortified her female frame with Drake's Plantation Bitters;
And, too nervous for severer exercise than waiting,
Let her student run the books and do the operating.

On the way Tom spent the day a planning out the meeting,
Setting to the letter e the items of their greeting;
How to clasp her tiny hand, around the neck to hold her,
While her dainty, downy cheek reposed upon his shoulder.

*The telegraphic contraction for "good morning."

What was his astonishment, when first he stood before her?
What was her's when first she faced her long, long time adorer?
His, to find her slim, and grim, and gaunt, and five eleven;
Hers, to see him old and fat, and barely four feet seven!

Cupid's dart might bring its smart e'en to this aged duffer;
Nancy Anna's spinster heart, though old and tough, could suffer.
Thus to meet and thus to part, was rough enough for certain;
Let us drop a briny—and by all means draw the curtain.

Who of you who read these lines, while plying the bandanna,
Recollects her Thomas Tot, or who his Nancy Anna?
Shall I pass a warning word to point my modest moral?
Pshaw! what dictum teaches babes there is no milk in coral?

MARRIED BY TELEGRAPH.

On the contrary, maids have been both wooed and won by telegraph, and in the year 1874, a minister married in the Keokuk, Iowa, office of the Western Union Telegraph Company, a couple at Bonaparte, in the same State, he performing the ceremony and they pronouncing the marriage vow over the wire. Five o'clock, April 16th, were the hour and the day fixed for the ceremony, and precisely at that time a dispatch was sent to Keokuk to the effect that the candidates were at the telegraph office in Bonaparte, and ready to proceed. The following was then sent:

"KEOKUK, Iowa, April 16th, 1874.

"JOHN SULLIVAN and FRANCES GODOWN,

Bonaparte, Ia.:

"Please join hands and take the pledge.

"WM. C. PRATT."

The following is a copy of the pledge which had been left with them.

"You mutually and solemnly promise before God and the witnesses present, that you will each take the one you hold by the hand to be your lawful and wedded companion. That, forsaking all others, you will cleave to each other in sickness and in health, and perform all the duties of a faithful companion until you are separated by death. If to this you agree, send me a message to this effect."

Then came the response:

"BONAPARTE, April 16th, 1874.
"WM. C. PRATT, Keokuk:

We take the pledge.

"JOHN SULLIVAN.
"FRANCES GODOWN."

The concluding dispatch was then sent as follows:

"KEOKUK, Ia., April 16th, 1874.
"JOHN SULLIVAN and FRANCES GODOWN,

Bonaparte, Ia.:

"By authority I pronounce you husband and wife, and may God bless you.

"WM. C. PRATT."

The operators all along the line then tendered their congratulations to the happy couple upon their marriage by the lightning process. Managers Dolbear, of Keokuk, and Detwiler, of Bonaparte, were the officiating telegraphists. This was the first marriage by telegraph, so far as there is any record. Several have been so celebrated since, and many more, doubtless, will be;

but we deprecate the insinuation which has been made, that divorces will be obtained by the same means.

HOW AN ABSCONDER WAS CAUGHT.

Now and then an operator proves himself unworthy of the profession, as did a young fellow named D. B. Leber, who, at the close of his telegraphic career, got into disgrace by stealing a package containing two hundred and fifty dollars, from the express agent at Watseka, Ill., where he was employed as operator. He also at the same time forwarded to the secretary of the telegraph company a package purporting to contain sixty-two dollars, but which was filled with blank paper. He then left by train for Chicago, calculating that as there was no other operator at Watseka, he would have time to effect his escape. But he was caught by means of another man there, whose knowledge of telegraphy was confined to making the alphabet, but who, upon the discovery of the theft, opened the key and sent a message three or four times, to nobody in particular, informing whoever it might concern that Leber had skedaddled with the cash, as above related. He could read nothing that was said to him, and continued to repeat his announcement at short intervals, even when other offices were engaged in sending messages on the wire, until a man was sent there by train to shut him off. His timely information, however, resulted in the capture of the thief, who was arrested in Chicago, and the money recovered.

The feat of this novice in the art is remindful of what

has been accomplished by defter fingers, if not apter intelligence, than his, in the way of

WONDERFUL SPEED IN TELEGRAPHING.

It is stated that no operator of modern times has been found to exceed the sending speed of Jo. Fisher, of Nashville, to Jimmy Leonard, of Louisville, in 1860 or 1861. The rate was an average of either fifty-three or fifty-four words a minute for ten consecutive minutes. The matter was press report. No better receiver than Mr. Leonard, who copied it, has yet been reported. A telegram was sent from London to Washington in nine minutes and thirty seconds. Two thousand five hundred and eighteen words were sent from New York to Cleveland in an hour. On the day of Mr. Lincoln's funeral, the American Telegraph office in Washington transmitted seventy-five thousand words of reports for newspapers in New York and elsewhere. All but about five thousand of the whole number of words transmitted were sent after 7 p. m., and it was all through at twenty minutes after 1 a. m., being at the rate of twelve thousand words per hour. Eight wires were in constant use, and nine part of the time. All this was accomplished in addition to the large amount of private business of the line. About ten thousand words of press news in addition were sent by the United States line, making a total of eighty-five thousand words sent to and paid for by the press of the country in one day from Washington alone, at an expense of about three thousand dollars. Thirteen thousand six hundred words were transmitted by the House printing instruments on a single wire after half-past 7 o'clock.

When, on one occasion, the lines were connected through from San Francisco, California, to Heart's Content, Newfoundland, the terminus of the Atlantic cable, after exchange of the usual complimentary messages, at twenty-one minutes past 7 A. M., Valentia time, a message was started from Valentia for San Francisco, passing through New York at thirty-five minutes past 2 P. M., New York time, and was received in San Francisco at twenty-one minutes past 11 P. M., San Francisco time, and its receipt at once acknowledged. The actual time occupied was only two minutes, and the distance traversed fourteen thousand miles, though the largest distance worked in one circuit was but five thousand miles, namely, from San Francisco to Heart's Content. Subsequently the operator at San Francisco transmitted an eighty-word message to Heart's Content direct, occupying three minutes in transmission, which was repeated back by the operator at Heart's Content in two minutes fifty seconds.

These wonderful accomplishments remind us of Shakspeare's gentle Puck, who, responding to an order from the fairy king, says:

> "I'll put a girdle round about the earth
> In forty minutes."

THE MESSENGER SERVICE

has been perfected in New York by the American District Telegraph Company, which employs nearly a thousand uniformed boys, none under fourteen years of age. They answer summonses at all hours, from over five thousand boxes, in dwelling-houses, stores, etc., in all parts of the city. The various uses to which the messengers are put are remarkable. Of late there has arisen a demand for escorts to places of amusement, and from one house to another. The former has become a regular practice. One evening recently there were eight ladies at six different theatres whose escorts were furnished "to order." Men as well as women employ escorts for various purposes. Most people who require the services of the messengers are strangers, who wish for guides to show them the "sights." Another use that is made of the District Telegraph messengers is to attend children, particularly girls, to and from school. Cases are not unknown where a messenger has been summoned and sent in search of a missing husband, who was supposed to be at one of his favorite haunts. It is not an uncommon thing for a messenger to be sent home with an intoxicated person. Messenger boys and men are also extensively employed as detectives for various purposes. Special messengers, or men or boys in plain clothes, are assigned to special duty as "spotters" of suspected clerks in stores, and they are said to have done excellent work. In fact, detective duty appears peculiarly adapted to those in the messenger service. Another use which has been found for messenger boys is the paying by proxy of New Year's calls. They are also employed as ushers at fashionable weddings, and as "managers" of the arrangements for carriages on such occasions. The books of the company show the services for which the boys have been required, and many laughable records are to be seen. One boy was detailed to take care of a lady's poodle, for which he was paid thirty cents an hour. An escort was required to attend to the theatre a lady whose husband was to "come later." A young man was once telegraphed for in order to bring a bumptious servant to terms. During political campaigns the boys are employed extensively to distribute documents. Car-drivers, and, indeed, all classes of people who have to get up very early in the morning, are peculiarly dependent upon the messenger-boy system. The books also show that the messenger boys have been used to order dinners, to buy all kinds of liquors, to do shopping for women, to pay bills of all amounts, and even to borrow umbrellas. Not unfrequently boys are sent to pawn-brokers' shops with articles.

THE TELEGRAPH MESSENGER.

There is perhaps no person who sees more of the different phases of human nature than the messenger connected with the regular telegraph companies. He is hailed at one door with anxious, enthusiastic joy; at another with superstitious dread, and at another with an impatient nervousness, which has the effect of making the person to whom the telegram is addressed, snatch a leaf from the receipt book instead of tearing open the envelope of the doubtful message.

The messenger rings the door-bell of Mr. Jones' residence. Mr. Jones attends the call. On seeing the messenger present the telegram, he hurriedly tears off the wrapper and proceeds to read it over and over, and finally asks:

"Is this for me?"

"It seems to be your address."

"Where will the marriage ceremony be performed?"

"I don't know, sir, anything about it; please sign and let me go."

"Oh! it's a telegram! I must tell my wife." And the door abruptly closes in the face of the messenger.

On his route the messenger stops to deliver a dispatch to Mrs. Spilkins. The family are at dinner.

"Bridget, who rang?"

"Missis, it's for you." And Bridget hands her the telegram. A shriek, and she falls backward, her lips faintly murmuring "telegram!" After sufficiently recovering, she remarks: "I told you, Mr. Spilkins, about the dreadful dream Mrs. Smith had last week."

"Oh! poor Jane—when will the funeral take place?"

"And her poor children—oh! how can I bear it?"

"Mr. Spilkins, you wicked man, how can you smile while you read it?"

Mr. Spilkins commences reading aloud:

"Petersburg, December 1st."

"Oh! Mr. Spilkins, don't read it to me—"

He continues: "Dear Mother—"

"And did she write it before she died?"

"Dear mother, all well. I and the children will be over on the early train to-morrow."

"Ah! I knew it was no bad news; but I am always so nervous about a telegram."

A MESSENGER MISTAKEN FOR A POLICEMAN.

A story is told which suggests that the blue coats and decorations have led to the supposition that the wearers are policemen in miniature. A short time ago a boy was sent with a telegram for a son of the Emerald Isle, whose name was Mulligan. The woman of the house came to the door, in answer to his summons, and, seeing his uniform, surmised at once that her Pat had been cutting up some of his shines again, and resolved to save him from the lock-up at all hazards. "Does Patrick Mulligan live here?" "Indade, sir! me Pat was drafted into the army, an' sure an' he's gone way off, an' I don't know where he is, at all." "Well! here's a telegram for him." "A telegram! fhat's that?" "Why, it's a dispatch—a message." "Do yees mane a telegraf dispatch, something like a letther?" "That's it, exactly." "Is that all? Faith, an' if you'll be afther goin' over forninst the grocery ye'll find him there smoking his pipe on the stoop. I took yees for a cop."

Thus much of operators and their useful allies.

THE TELEGRAPH IN WAR.

In the introductory chapter we showed how, centuries before the Christian era, as dated in records which are considered authentic, signaling by fire was employed as a means of advantage in military operations, and that the comparatively clumsy signaling arrangements in use just previously to the introduction of the electric telegraph, had one of their principal occupations in communicating military doings and events.

In general, it may be stated that sun-signaling, which is, of course, only practicable in day-time, has advantages over all other methods of visual telegraphy. Messages can be transmitted to great distances, and the clearness with which the signals can be made renders background of but little importance, while in flag-signaling the distinctness of the signal depends materially on this question.

At the present time all the armies of the civilized world are provided, while engaged in actual campaigning, with a field telegraphic system, more or less efficient, besides availing themselves of local and existent means of lightning communication wherever practicable. Telegraphy in war was never employed to equal advantage and with greater perfection than by the Prussians in the campaign of 1870-1; but in our own country, the world witnessed its most gigantic operation.

Before giving an account, and a necessarily brief one,

of the extent and value of the telegraph in the American civil war, a general sketch of its employment in military operations may not be out of place.

FIELD TELEGRAPHY.

The English army, it is said, was the first to use it. In the Crimean war their trenches and batteries before Sebastopol were traversed and connected by lines of telegraph. The French soon followed their example, and constructed a similar system in their own lines, while, later on, a cable laid across the Black Sea put the armies in the field in direct communication with Paris and London. Since that time a regular telegraph corps has been organized in every European army. The field telegraph was used by the French in Italy in 1859, and in their campaigns against the Kabyles in Algeria; and in America both the Federals and Confederates made free use of permanent and temporary lines during the War of Secession, the Southern cavalry, in particular, displaying great daring and enterprise in riding round the flanks of the Federal armies, seizing their telegraph lines, sending false messages to the Northern generals, and then cutting the line and retiring as rapidly and secretly as they came. It was, however, as before stated, in the Prussian army, and in the great campaigns of 1864, 1866 and 1870–71 that military telegraphy attained its greatest development; and after the experience of these three wars, the Prussian telegraph corps is probably the most efficient in Europe.

The object of the field telegraph is to keep the headquarters of an army in communication with its several corps, and at the same time with the general telegraph system of the country. The line may be either an aerial or a ground wire, or a combination of both, the former being stretched on poles, while the latter is insulated by being enclosed in a light cable, about half an inch thick, and laid along by the roadsides or across the fields.

Where there is an extensive telegraph system in operation, all that is necessary is to connect the head-quarters of the army with the nearest point on a permanent telegraph line, and in most European countries any army in the field would seldom, if ever, be more than ten miles from such a line. Ten miles of the field telegraph can easily be erected in half a day; indeed, the Austrian engineers assert that on favorable ground they could do the work in two hours. In most cases, of course, the advancing army would have to repair the permanent lines which would be partially destroyed by the retreating forces, and in this way twenty-five miles of wire were often erected by the Prussians in a single day. As soon as an army moves forward, the field telegraph line previously erected is taken down, while a fresh line is laid from the new head-quarters to the nearest permanent telegraph. This is done with a view to economizing the material, an enormous amount of which would have to be carried with the army, if the lines it left behind it in its advance were not removed, and the poles, wire and insulators employed in their construction again utilized.

The conducting wires of the military telegraphs

which are used by the French army, are so made as to be capable of resisting the trampling of horses and the crushing of wheels of the heaviest vehicles on common roads, though not that of artillery or of a railway train.

INTERRUPTIONS AND WIRE "TAPPING" BY THE ENEMY.

While the field telegraph affords a commander a rapid and certain medium of communication with his base of operations and the various corps of his army, it must be remembered that it is one which is continually liable to interruption by an enterprising enemy. Wherever a general has to contend with an army well provided with good cavalry, he will find it extremely difficult to protect his telegraph lines from being destroyed by daring raids of his opponents. There are several easy ways of making a telegraph line temporarily useless. The simplest and most obvious method is to pull down the poles and cut the wires into pieces; but when this is done the damage is easily detected, and the repairs at once commenced. The interruption will, therefore, be far more serious if it can be effected in a way which will not permit of its exact locality being so readily discovered. This can be done by cutting the wire, introducing a piece of gutta percha or any other non-conducting substance into the course of the circuit, and connecting the ends of the wire with it, so as to give it the appearance of one of the ordinary joints or splices of the line. At the same time a few poles can be pulled down in another place, and the wires cut, and the probability is that the engineers who repair the line will not discover the hidden interruption of the circuit until after they have restored the gap, and found that the wire is still cut somewhere else; and even then the place where the non-conducting substance is introduced will not be discovered until some time has been employed in carefully testing the line with the galvanometer.

But there are other dangers to telegraphic communication in the field besides the mere damage to the line. If the enemy's cavalry get possession of a station, they can easily send messages containing false information or delusive orders to well-known officers of the opposing force, while the place from which they are sent and the assumed name in which they are dispatched, will give the messages an appearance of authenticity which, if it does not completely deceive the recipient, will at least be the cause of considerable doubt and perplexity to him, and, perhaps, make him hesitate to accept the accurate information or authentic orders received from other sources. Again—even without occupying a station it is possible to read the messages which are passing along a telegraph line, and thus perhaps discover important secrets. All that is required for this purpose is a small portable receiving instrument and a few yards of copper wire to connect it with the line. A single individual thus equipped can "tap" a telegraph line and read whatever messages may be passing over it.

These dangers, however, are only of a partial or temporary character. By carefully patrolling and testing the line, it cannot be interrupted for any length of time without the damage being observed and repaired. By adopting a secret arrangement that there shall be a certain number of letters in the two or three words at the beginning or end of every message, a dispatch sent by an enemy can in most cases be detected. And, again, by employing a cipher alphabet, it will be difficult for any one who taps the line to obtain information from the messages which fall into his hands.

FIRING GUNS BY ELECTRICITY.

Electricity is now applied in the firing of artillery, an improvement introduced by Mr. M'Kinlay, at Woolwich, England, in the year 1856, when the "galvanic tube" was invented. In this tube a steel or platinum wire is embedded in a charge of powder, and this wire forms a link in the circuit of a galvanic battery. The retardation of the current, due to the inferior conducting power of the steel or platinum wire, causes it to be raised to a red heat, and by this means the powder is exploded. This system was in use until 1862, when the Abel "electric tube" was invented. In this the steel wire is replaced by a priming charge, consisting of subphosphide and subsulphide of copper, with a little chlorate of potash, and in this composition the terminals of the two insulated copper wires that conduct the etectric current are embedded. The points of the wires are about one sixteenth of an inch apart.

A later innovation in military matters is the introduction of the electric light for the purpose of illuminating camps, which has been successfully adopted by English volunteers.

Experiments recently made go to show that the telephone will probably also prove a valuable adjunct in military operations.

But to our civil war, which affords much interesting material whose insertion is forbidden by lack of space.

THE CIVIL WAR.

General Sherman has written: "For the rapid transmission of orders in an army covering a large space of ground, the magnetic telegraph is far the best, though usually the paper and pencil, with good mounted orderlies, answer every purpose. I have little faith in the signal service by flags and torches (though we always used them), because almost invariably when they were most needed the view was cut off by intervening trees or by mists and fogs. There was one notable instance in my experience, however, when the signal flags carried a message of vital importance over the heads of Hood's army, which had interposed between me and Altoona and broken the telegraph wires—as recorded in my 'Recollections;' but the value of the magnetic telegraph in war cannot be exaggerated, as was illustrated by the perfect concert of action between the armies in Virginia and in Georgia in all 1864. Hardly a day intervened when General Grant did not know the exact state of facts with me, more than 1,500 miles off as the wires ran. On the field a thin insulated wire may be run on improvised stakes, or from tree to tree, for six or more miles in a couple of hours, and I have seen operators so skillful that by cutting the wire they would receive a message from a distant station with their tongues. As a matter of course the ordinary commercial wires along the railways form the usual telegraph lines for an army, and these are easily repaired and extended as the army advances, but each army and wing should have a small corps of skilled men to put up the field wire and take it down when done. This is far better than the signal flags and torches. Our commercial telegraph lines will always supply for war enough skillful operators."

ORIGIN OF THE U. S. MILITARY TELEGRAPH.

On the occasion of the riots in Baltimore, April 19th, 1861, the rebels, by destroying railroads, burning bridges, and tearing down lines of telegraph, succeeded in cutting off all communication between Washington and the loyal States.

The object was to prevent reinforcements from reaching Washington, so that the rebel leaders might concentrate their forces on the banks of the Potomac and demand the surrender of the Capital before the Government could summon sufficient aid to its defense.

The work of rebuilding the destroyed property was intrusted to Colonel Thomas A. Scott, the well-known late president of the Pennsylvania Railroad. Under his direction the rails were soon relaid, bridges rebuilt, and new telegraph wire erected.

A party of four telegraph operators was organized in Pennsylvania, April 25th, 1861, and at once started for Washington, which city they reached by a circuitous route on the 27th. This quartet formed the nucleus of the United States Military Telegraph, many operators from different parts of the country being afterward added.

COST OF THE SERVICE DURING THE WAR.

During the Rebellion there were constructed and operated about 15,000 miles of military telegraph. The cost of the service from May 1st, 1861, to Dec. 1, 1862. was about $22,000 per month. During the year 1863 it averaged $38,500 per month. In 1864, the telegraph was greatly extended, and the cost reached $93,500 per month. The total expenditure during the year ending June 30th, 1865, was $1,360,000; and the total expenditure from May 1st, 1861, to June 30th, 1865, footed up $2,655,500.

THE DUTIES OF CIPHER OPERATORS.

Throughout the war the cipher operators connected with the United States Military Telegraph, under Generals Eckert and Stager, were at all army headquarters. Their duties were confidential and very important, inasmuch as all military movements ordered by General Grant were transmitted through them. They were in possession of intended army and navy expeditions sometimes weeks before commenced, and, had they not been patriotic and truly loyal, could have defeated the Union armies and delayed their final triumph. These quiet, unassuming gentlemen were very poorly paid, and frequently not well provided for. However, they did not complain of their hardships, but worked on faithfully until the Rebellion was crushed. The operators, it may be added, were not commissioned, nor even borne on the army rolls, and having no discharges from the service, will not be remembered by the country and their valuable services acknowledged like officers and soldiers. They, however, did their duty nobly, as did also the operators employed in less responsible positions. The telegraphic service employed in the war received some official recognition of their patriotic services in the honor done a few of their representative men, in accordance with the following communication, which speaks for itself:

ACKNOWLEDGMENT OF MERITORIOUS SERVICES RENDERED THE GOVERNMENT.

OFFICE U. S. MILITARY TELEGRAPH,
WAR DEPARTMENT, WASHINGTON,
July 31, 1866.

D. H. BATES, assistant manager department of the Potomac.
Charles A. Tinker, chief operator, war department.
Albert B. Chandler, cipher and disbursing clerk, war department.
A. H. Caldwell, chief operator, army of the Potomac.
Dennis Doren, superintendent of construction, department of the Potomac.
Frank Stewart, cipher clerk, war department.
George W. Baldwin, cipher clerk, war department.
Richard O'Brien, chief operator, department of North Carolina.
George D. Sheldon, chief operator, Fortress Monroe, Va.

M. V. B. Buell, chief operator, Delaware and Eastern shore Line.

John H. Emerick, chief operator, army of the James.

GENTLEMEN:—I have been instructed by the secretary of war to present to each of you one of the silver watches which were purchased and used to establish uniform time in the army of the Potomac, marked "U. S. Military Telegraph," as an acknowledgment of the meritorious and valuable services you have rendered to the government during the war, while under my direction, as an employe of the United States Military Telegraph.

It gives me great pleasure to comply with these instructions, and I take this occasion to thank you, for myself, for your faithful performance of the important trusts which have been confided to you in the various capacities in which you have served, and especially as "cipher operators."

Yours, very truly,

THOMAS T. ECKERT,

Ass't Sec'y of War, and Sup't U. S. Mil. Tel.

AN OPERATOR'S READY WIT.

Instances well nigh innumerable could be given of the ingenuity manifested by operators under circumstances of danger, and which, as in the one here cited, proved of great value to the patriotic cause. When the rebel General Morgan made his great raid through Indiana and Ohio he captured a Union operator, and compelled him to telegraph, in General Lew Wallace's name, to Cincinnati, asking how many regular troops were in that city. Morgan read by "sound," and therefore the operator did not dare to intimate that he was under duress, and could only venture to add an extra initial to his own signature. The receiving operator at Cincinnati knew that Morgan was in that neighborhood, and suspecting, from the extra initial letter, that all was not right, replied, greatly exaggerating the force of regulars; and the consequence was that Morgan changed his route to a circuit of twenty miles beyond the city, and thus saved it from a sack, and the probable loss of millions of dollars.

HEROIC COURAGE OF AN OPERATOR.

Great Falls was a Union picket post, where Federal troops watched rebel movements on the Virginia side of the Potomac. The well-known telegrapher, since deceased, Ed. Conway, a Canadian, was government operator. One afternoon the United States pickets were withdrawn. The rebels thought it was a good opportunity to try the range of their guns; so, coming in, a considerable number of them began to fire away at the telegraph building, wherein Conway was bewailing the condition of his finances. Shells flew thick and fast around the building—steps and porch were soon blown away, but the plucky telegrapher heeded it not. They mixed a volley of musketry with the firing of shell, but this only caused him to gather up his three cents and a button, place them in his pocket, and whistle "Johnny went for a soger." A quantity of bullets came unceremoniously into his room, and as unceremoniously as they had come in he went to work digging them out of the partition, to be saved as trophies. Only when the rebels began crossing the river did he consider it worth while to seek other quarters. Such courage has rarely been equalled, even by men accustomed to the vicissitudes of war.

NERVOUS OPERATORS.

On the night of May 23d, 1861, the night before the occupation of Alexandria by the Federal troops, the Union operators at the Chain Bridge, Woodhouse and Jacques, seeing a great stir among the soldiery, imagined at once that preparations were being made for a retreat instead of a victorious advance, and at once telegraphed to Mr. Strouse, the superintendent, that something was up, and, fearing a retreat, they had no means of escape unless he immediately sent them two horses. Danger was at hand, and he alone could protect them from an infuriated enemy; their lives should be at his disposal if the necessary protection was forthcoming. The horses were not sent, however, nor were their lives sacrificed; on the contrary, they are living to-day.

A FUNNY WAR STORY.

"Agitator" told a good story in *The Telegrapher*, how that during the early part of the month of November, 1863, General Sherman, then commanding the 15th

Army Corps, was making a forced march across the country from Memphis to Chattanooga, Tenn., to support Gen. Rosecrans, who had been partially defeated at Stone River. Upon reaching Elk River the telegraph and cipher operator attached to General Sherman's staff received orders to proceed to Decherd, Tenn., the nearest telegraph office, seventy-five miles distant; send important military dispatches to General Grant at Chattanooga, receive replies, and hasten back to meet the corps' advance. One hundred of the 3d regular cavalry were detached as an escort, and on the third of November set forth. As this mission was important, no time was lost on the march, although the roads were in a terribly muddy condition, and great caution had to be observed against surprise by Confederate bushwhackers. Fast riding and muddy roads do not add much to the outward appearance of man or beast, and by the time Decherd was reached the staff operator presented about as sorry an appearance as could well be imagined.

The operator pulls up in front of the telegraph office about four o'clock one very rainy afternoon. Entering, he is greeted with the familiar click. There, in a little eight by ten pen, laboriously at work trying to "break" some obstinate "plug," less experienced in telegraphy than himself, sits that nervous, mischievous little sprite, Jimmy Lowe, the operator. Jimmy is not in the best of humor at this particular time, and dislikes to be interrupted "when in for a fight."

Thinking he is a student, our friend inquires if the operator is in, accompanying the inquiry by an awkward movement. "Yes, I am the operator. What do you want?"

Now, the chance for a good practical joke could not be resisted by our horseman, therefore he quickly decides to have a little fun at Jimmy's expense.

"What sort of a clicking affair is that 'ar?" he enquired, pointing to the register, with its ponderous weight and paper tape.

"That is the telegraph," says Jimmy, "and I am the operator. Do you want to send a message? If not, don't bother me, but go and get some of that mud off from you." Jimmy turns away with a look of disgust, and proceeds to renew his battle over the wire. It will here be proper to state that Jimmy kept a sutler's stand on a small scale in one corner of the office, and, as he afterward acknowledged, was suspicious that our friend had an eye on a quantity of plug tobacco behind the counter.

After a great many questions relative to the *modus operandi*, all of which worked Jimmy's nerves up to a perceptible tremble, our staff man concluded to bring the matter to a focus.

"See here, stranger, p'rhaps I kin help yer. Just let me in thar, will yer? That tarnel clatter has been agoing on long enough. You won't, eh?" With one stride he clears the board railing and brings up by Jimmy's side, with open mouth gaping at the instrument.

Jimmy is stormed in his stronghold; confounded, and not knowing what to say, prudently says nothing. He, however, involuntarily drops his hold of the key, and has half a mind to close in with his muddy tormentor, but does not. Mr. Cavalry-man sidles around and gets hold of the key. Jimmy is now nearly frantic; visions of Confederates in disguise flit through his mind, and he looks around for chances of escape. He can read just enough by sound to know that our friend has given a signal for precedence over the wire. He hears him "call" Chattanooga; he hears Ch. answer. Oh, if he could only get hold of the key now and warn Ch. of danger. He knows our muddy friend is a Confederate operator in disguise. It is now his turn to stare with gaping mouth.

Our friend coolly transmits the dispatches, politely calls for pen, ink and blanks, and receives the long replies without a break, and without using the paper tape. Jimmy cannot make out the purport of what is going on over the wire, and our friend, by hiding the blanks with his hand while receiving, keeps him in the dark. All is soon finished. The dispatches are folded, placed in an inner pocket; and with many thanks for the courtesy extended, our friend retires from the office, mounts his stalwart steed, and is soon cantering off to meet his general. It afterward came to light that Decherd asked Chattanooga some queer questions over the wire soon after the "raid."

ANOTHER,

which came from Jefferson City, Missouri, tells that during the reign of terror in the county distinguished by including that city within its limits, caused by Price's raid, the depot at Jefferson City caught fire and was burned. Consequently, the operator was obliged to find other rooms for the telegraph office, and, for want of better, located temporarily in Dad Chevron's carpenter shop. One day, during the absence of the operator, all the instruments commenced and for fifteen minutes kept up a terrible ticking, which frightened the old man, who had not made the science of electricity the great study of his life. He thought it must be a call for his office, and probably conveyed news of Price and his forces. Making a dive for one of the instruments, he caught the "ground-wire" firmly between his teeth, and shrieked out: "Operator's gone to dinner; be back in half an hour!" and at the same instant received a shock from the wire coming into contact with his moist tongue that he will remember to his dying day.

POOR QUARTERS FOR TELEGRAPHERS.

When Operators Lathrop and Maize went to open the office at Langleys, Va., for the use of General Smith's division, they found that no provision had been made for them, and accordingly went to General Smith to have him point out their location. The general eyed them for a few minutes with a scrutiny worthy of a Bow street detective, and then made a reconnoissance for a proper location. After a faithful survey of half an hour, he espied an old shed, raised some feet from the ground, in the basement of which were some horses, cows and pigs, and above these a room in which the cook kept his poultry. This latter apartment he ordered to be divided by a partition, one side to be occupied by the operators, the other for poultry.

A PROVIDENT OPERATOR.

During the war operators suffered their share of

the discomforts and hardships incident to campaigning. The young gentleman immortalized in the following story had an eye to future necessities in his preparations; and who will blame him?

Two days after the battle of Ball's Bluff (October 21, 1861), wherein the Union troops were repulsed and Colonel E. D. Baker, the patriotic senator, was killed, a telegraph office was opened at Edward's Ferry, on the upper Potomac. Mr. Tinker and a white-haired youth named Burnker were sent to the office to "do" the telegraphic honors. The office was located in a hut, vacated but a short time prior to their arrival by a lame contraband. Two days only did the office remain at the Ferry, for no sooner did General Banks, with his column, leave there and return to his former headquarters, near Darnestown, Maryland, than the boys thought that "discretion was the better part of valor," and, turning their backs to the enemy, pulled up traps and made the best time on record in rejoining General Banks.

Burnker, when preparing to leave home to join the U. S. Military Telegraph, wisely foresaw that circumstances might occasion his being sent to some distant camp where forage for telegraphers would not be supplied by the quartermasters, and filled a trunk with eatables. Arriving in Washington, he found it necessary to make his way to Edward's Ferry, and, having no means for their transport, to leave his commissary stores behind him. Being absent ten days, and no prospect of being able to have his provision sent him, he requested one of the operators in Washington to open the trunk and take care of the contents. The trunk being opened, disclosed to the gaze of the hungry opener seven pounds of pound-cake, six pounds of fruit-cake, one peck of apples, half a bushel of chestnuts, a bologna sausage, a head of cabbage, and six turnips. On the fact becoming known that he was such an excellent provider, the position of quartermaster and commissary of the corps was tendered him; but, possessing the modesty for which all telegraphers are proverbial, he respectfully declined the position.

THE END—RICHMOND TAKEN.

No message ever sent by telegraph was of so much national interest as the one which William E. Kettles, an operator in the service of the government at the war department in Washington (at the present writing on the staff of the Boston, Mass., Western Union main office), received from Richmond on the morning of the 3d of April, 1865. Mr. Kettles, then a mere boy of fifteen, was working the Fortress Monroe and City Point wire at Washington. Shortly after 9.30, the Washington and Cherrystone operators were engaged on a long message, when suddenly both men were taken aback by what seemed to be a most foolish demand from Fortress Monroe: "Turn down for Richmond, quick!" Had a flash of lightning struck through the walls at that moment, the shock could not have been greater than it was on the part of every man in the room.

There was great alacrity in turning down the adjustment. There were trembling fingers while it was being done, and there was a gathering around of many operators, with curiosity, suspense, and impatience combined, to see what it meant. Sure enough! the signals from the operator in Richmond to the operator in Washington were bounding along the line. No signal was ever answered more promptly. Then came the question:

"Do you get me well?"
"I do; go ahead!"
"All right. Here's the first message for four years:"

"RICHMOND, VA., April 3d, 1865.
"*Hon. E. M. Stanton, Secretary of War:*
"We entered Richmond at 8 o'clock this morning.
"G. WEITZEL,
"Brigadier-General Commanding.."

Mr. Kettles concedes that he copied the message, but he could never tell how. He remembered starting up from his chair and upsetting inkstand and instrument; of kicking over a tin that sat at the fire-place, in order to make a noise; of rushing for General Eckert's room, where sat President Lincoln and Mr. Tinker, the cipher clerk, talking in a low tone. As Kettles was about to hand the message to Mr. Tinker, the President caught sight of the body words, and, with one motion and two strides, message and President were out of sight on the way to Secretary Stanton's room. Mr. Tinker and everybody else were dumbfounded. Kettles quietly returned to his instrument, walking like one in a dream; proceeded mechanically to turn the inkstand right side up, and to straighten up his overturned machine. Then he sat down in his chair, and stared before him in blank amazement. Around him were the other operators, every man alike flustered, and unable to get their minds back to their work, or to utter connected words.

In less than one-quarter of the time it takes to write this, the operating room was filled with officers and sub-officers. President Lincoln and Secretary Stanton came in and shook hands with every one in the room, and then every one in the room shook hands with one another, and then with the President and Secretary again. Then they all crowded around the Fortress Monroe instrument, hungry for more news. Kettles sat at his instrument while questions were showered in on him from every mouth. He was asked more questions in those ten minutes than he will be likely to be ever asked again in that space of time. At last the information came that Richmond had disconnected itself for the present. All retired to General Eckert's room except Mr. Tinker and Kettles, who stood by the window endeavoring to hear themselves think. Neither of them had drawn a perceptible breath for ten minutes. Outside were the broad grounds of the department buildings. Looking from the operating-room window the prospect was clear; not a single person was to be seen. Suddenly a Georgetown horse-car appeared in the distance. On it came at the usual rate. Near the building it stopped. A man got off, and started with slow, leisurely steps up the center walk to the door. Inside the operating room the thrilled operators looked out on his slow, steady pace, and could scarcely contain themselves at his unconcern. He was meditating—actually meditating—as though there was nothing to throw off his hat for and cheer till he was hoarse. Keeping on, he presently lifted his head and looked at the window. Tinker was there and knew him.

"Any news?" he casually inquired.

Tinker leaned far out of the window: "Richmond's fallen!" he said.

No tongue can describe the features of that man while he was coming to himself. He turned red and white by turns, till, suddenly realizing the meaning of the words, he waved his arms, then turned and ran. Down the street he ran, spreading the news to every one he met. Soon there was a great crowd. The excitement rose; the people seemed almost wild. The War Department was soon besieged. Outside was a multitude. Inside were excited officers, clerks, operators, and an excited President. The outsiders looked in at the insiders, and the insiders looked out at the outsiders. Questions came hot and fast from the multitude, and answers were shouted back from every man who could get his head to one of the two windows. The crowd got the news fairly in its mind and then seemed to want three cheers. The three became four. Then they wanted speeches. They got them. Half a dozen speeches were under way in less than that many minutes. Some were good ones. Andrew Johnson was there. He was saying: "God bless the old flag! If I was President of the United States——"

At this point something exciting occurred. Secretary Stanton entered the operating-room leaning on General Eckert's arm. General Eckert pointed out to him the boy who had received the message. They were formally introduced. The next moment Kettles found himself seized by the secretary and held at arm's length out of the window above the crowd. The secretary called to the crowd that this was the young man who had received the dispatch of the fall of Richmond. The crowd wanted a speech from him. Kettles gave them a speech in a few words, appropriate and pointed, for he was in the humor.

Then followed other scenes. Fire-engines were brought out—anything to make a noise. In the evening the city was ablaze with illuminations. Kettles, who is now an operator in Boston, says he can never forget how Father Abraham started for Secretary Stanton's door after receiving the dispatch—hop, skip, and jump—shouting: "Clear the track!"

THE ASSASSINATION.

Joy, however, was speedily turned into mourning. A writer in the Albany *Evening Journal* eloquently tells the story of the great crime of April 14th, 1865, which plunged the nation into grief unutterable:

"One calm night in the spring time, when the silver stars were gleaming out pensively, and scarcely a footfall on the pavement of Broadway or State street broke the stillness that reigned, the cupola-man on the City Hall had intoned the midnight hour, and added: 'All's well,' when a sudden, nervous call of 'rep, rep,' aroused all on the line from Washington to the red man's home in the far west, and to the southwest, where the green grass waved in luxuriance, and the little birds twittered their matin songs from among the boughs of blossoming trees, as well as to the icy fastnesses of Halifax and the Canadas—to all alike came the harrowing words: "'Tis rumored that the President was shot at the theater to-night!' How our hearts seemed rent asunder, and the great tears swelled up to the eyes that for years previous were strangers to such outward expressions of sorrow. Soon after another message came, saying: 'Suppress that rumor sent you; it's all false.' What muttered threats and words followed one another over the wire to headquarters after the reception of this latter will never be known but to those in attendance that sad and fatal night. Again all was quiet, and the clock ticked away the moments, and the hands sped around to the morning hours, when 'rep,' rep' was again sounded, and the brass instruments clicked out an 'official,' giving the whole dark and bloody tragedy of the assassination of the lamented President. Sad and wan was the face of our little report-boy, 'Patsey,' as he handed into the offices of the morning papers the heart-rending account; and nervously the hands of the weary compositors picked up the letters that, set into form, recorded the assassination of President Lincoln on the evening of April 14th, 1865."

THE FRANCO-GERMAN WAR.

We shall conclude this chapter with two reminiscences of the struggle of 1870–1, the first reflecting glory upon the Prussian arms and, we fear, some discredit upon our Gallic friends, but the latter redeeming this disgrace by the heroism of a French female operator.

At Manheim there was lately on exhibition a telegraphic apparatus taken from the French, which was obtained in the following manner: A certain dragoon of the Baden Guards, by name Muench, with two of his comrades, was sent to reconnoitre as far as the Vosges. On their entering the village of Raon l'Etampe the inhabitants fled in every direction, with cries of "The Prussians! the Prussians!" and shut themselves up in their houses. Thus left masters of the town, the dragoons rode to the town hall and summoned the mayor. They asked him where the telegraphic bureau was located. He pointed it out, and they went to it, and Muench, singly, and in the presence of the assembled city council, cut the wires, unscrewed the apparatus, and buckled it to his saddle.

The French government has recently conferred the military medal upon a young woman named Mdlle Dodu, employed in the telegraph office at Pithiviers during the war of 1870. Upon the arrival of the German forces in that town they at once, as was their wont, took possession of the telegraph office, and relegated Mdlle Dodu, who was in charge, to a room on the first floor. The wires passed through this room, and Mdlle Dodu managed to tap them and convey the information transmitted over them to the sub-prefect. In this way she kept the French military authorities cognizant of the designs and movements of the enemy.

CABLE TELEGRAPHS.

We have now come to the greatest triumph, so far, in ocean telegraphy—the connecting of the old and new worlds by cable, an account of which it may be well to precede with a few general remarks on ocean telegraphy.

Previously to the accomplishment of this undertaking, which, it will be remembered, was not successful in the earlier attempts, submarine telegraph cables had been laid and worked, but they were of comparatively little length.

THE ATLANTIC CABLE

is the one which most interests us Americans, and whose importance in business and the affairs of nations cannot be over-estimated, especially since duplex working has become an accomplished fact.

THE FIRST SUGGESTION OF AN ATLANTIC TELEGRAPH.

An old periodical contains the following paragraph, which is given here as embodying the first idea of telegraphic communication between Europe and this continent.

"Mr. J. B. Lindsay, of Dundee, who is at present in Glasgow, propounds a startling theory, that of forming an electric telegraph betwixt Great Britain and America, without employing submerged wires, or wires of any kind At a meeting in the Athenæum Mr. Lindsay illustrated his method. A large trough of salt water was employed, across which he transmitted the electric current, without any metallic conductor, the water itself being the only medium of communication. Mr. Lindsay explained that he had obtained similar results over a breadth of sixty feet of water. Some calculations have been made in regard to the expense, and Mr. Lindsay computes, according to his present information, that the cost of the necessary battery and land wires to establish a communication between England and America would not exceed sixty thousand pounds ($300,000)."

THE ORIGIN OF THE ATLANTIC CABLE.

The Atlantic cable is said to have originated with Cyrus W. Field, and was suggested to him in this way: A Roman Catholic bishop of St. Johns. Newfoundland, Bishop Muloch, advanced the idea that a line be built connecting St. John with the mainland, and then running a line of fast steamers to the west coast of Ireland, thus bringing America within a week of Europe.

In 1852, a Mr. F. N. Gisborne, acting upon this suggestion, commenced the erection of a line from St. Johns, through four hundred miles of dense forests to Cape Ray, there to connect with the inland lines. The following year, however, a short cable which he had laid gave out, and those who had invested money in the concern withheld further support.

Work had therefore to be suspended. In 1854, Mr. Gisborne came to New York, and made the acquaintance of Cyrus W. Field, who was much interested in the enterprise.

While studying this subject, and turning over the globe in his library, the idea flashed across Mr. Field's mind: "Why not carry the line across the ocean?" He went to St. Johns, Newfoundland, in March, 1854, and obtained from the legislature of that colony a charter granting an exclusive right for fifty years to establish a telegraph from the continent of America to Newfoundland, and thence to Europe.

THE ORGANIZATION OF THE FIRST ATLANTIC CABLE COMPANY.

On March 10, 1854, articles of association were signed. A company of five gentlemen sat in Mr. Field's parlor in Gramercy Park, and entered into the project. They were Peter Cooper, Moses Taylor, Marshall O. Roberts, Chandler White, and Cyrus W. Field. Peter Cooper became president of the association. Mr. White subsequently died, and Wilson G. Hunt took his place. Mr. Roberts died on September 11th, 1880. The association was called "The New York, Newfoundland and London Telegraph Company." The company's capital was $1,500,000, of which Mr. Field subscribed one fourth. A grant of £50,000 to aid the work was secured, as well as fifty square miles of public land, with a further grant of fifty more when the cable was laid.

It took more than two years to build the land line across Newfoundland and Cape Breton Island. While this was being done Mr. Field went to Europe and ordered a submarine cable, to connect Cape Ray and Cape Breton. This was sent out in 1855, and was lost in a gale in an attempt to lay it across the Gulf of St. Lawrence. The attempt was successfully renewed in 1856. This cable cost $1,000,000.

LAYING THE CABLE.

In that year Mr. Field again went to London and organized the Atlantic Telegraph Company, to carry the line across the ocean. He secured from the British and American governments aid in ships, and accompanied the expeditions which sailed from England in 1857 and 1858 for the purpose of laying the cable across the Atlantic Ocean. Twice the attempt failed, in 1857 and again in 1858. The third attempt proved successful, and in 1858 telegraphic communication was established between England and America.

In forming the cable, the great object was to combine lightness with strength. A single strand is capable of sustaining a weight of fifteen hundred and fifty pounds. The centrifugal force of the cable when paying out had to be carefully guarded against. The cable issued out of the tanks at the rate of six miles per hour, and was paid out by means of a brake drum. At the end of the blocks weights were suspended, on the regulating of which the perfection of the paying out depended. By watching the distance the opposite weights were suspended the strain upon the cable was ascertained. Water offered a great resistance to the cable. If the cable was light it would descend in an inclined manner; if bulky, then it would lie horizontally. The cable was three hours before it reached the bottom, and not before seventeen miles had been paid out. If the cable, when in the process of picking up, were drawn in straight line, it would snap; hence the utility of laying it slackly.

THE FIRST MESSAGE TRANSMITTED.

The first message sent over the Atlantic cable was the announcement of the death of James Eddy, "the first and best telegrapher in the United States," as the the dispatch published in the *Times* said. So incredulous were the public, that doubts were expressed of the genuineness of the news transmitted, and only when a dispatch conveying the action of Parliament on an important public matter was verified by mail two weeks afterward, were these dispatches accepted as real.

In a few weeks the cable ceased to work, and this on the very day that had been set apart in the United States as a day of thanksgiving for its completion. Although it was again pronounced a failure, Mr. Field never lost faith, and made frequent trips to Europe to resuscitate the company. The civil war broke out in the meantime, and not until 1865 was another expedition prepared. Submarine telegraphy had been greatly improved, a better cable was prepared, and the steamship *Great Eastern* took it on board, and sailed for the American coast. Over twelve hundred miles of cable had been laid, when, by a sudden lurch of the vessel, the cable snapped and was lost. The bottom of the sea was dragged for days in search of the broken end, and the expedition returned to England. In 1866, the *Great Eastern* again sailed with a fresh cable, and two thousand miles were safely stretched across the ocean, and the communication perfected July 27, 1866. After landing this the *Great Eastern* returned to the middle of the ocean, and after two months' search succeeded in grappling the sundered cable of the year previous. It was brought up from a distance of two miles, joined to the cable on the steamship, and carried safely to the Western shore. A weekly newspaper, called the *Atlantic Telegraph*, was published on the *Great Eastern* during these operations.

An amusing story is told of a gentleman who beset Mr. Field at the time when the cable was sundered with a proposal as to the manner in which it could be best raised from the ocean. This was to sink a hollow tube in which to go down and seek after the cable. Mr. Field was so annoyed by the continued calls at his hotel that one morning he told his visitor that it should be done, and that the author of the idea should make the first attempt. He never afterward saw the gentleman.

After twelve years of incessant labor, in which he crossed the ocean nearly fifty times, Mr. Field saw the crowning effort of his life accomplished. Congress voted him a gold medal with the thanks of the nation, and the prime minister of England said that it was only the fact that he was a citizen of another country that prevented him receiving high honors from the British government.

COST OF THE FIRST CABLE.

The first cable cost $1,256,250, and the company's expenditures up to December 1, 1858, amounted to $1,834,500. Among the dispatches sent over the cable was the speech of the king of Prussia just before the Austrian war. It cost $3,600 to transmit it. This cable has been in running order almost continually since its successful completion. In 1874 work was begun by the "Direct Cable Company" to lay a cable between Ballinskeligs Bay, Ireland, and Rye, New Hampshire, by way of Nova Scotia. It was completed in 1875. "The Compagnie Francaise du Telegraph de Paris a New York" completed, December 15th, 1879, the laying of a cable from Brest, France, to St. Pierre, Miquelon, and thence to North Eastham, Massachusetts, and an additional cable was laid by the Anglo-American Company in July, 1880.

RECENT REMARKABLE IMPROVEMENTS IN CABLE LAYING.

The rapidity with which the later cables, particularly the last two, were laid, is in striking contrast with the laying of the earlier trans-Atlantic cables. When the first attempts were made, the practicability of the scheme appeared doubtful. Two failures occurred.

The first cable, 1857–8, was defective, and, although between August 13th and September 1st, 1858, four hundred messages were sent between Valentia and the Newfoundland coast, yet the rate of reception was very variable, the signals often unintelligible and requiring repetitions. After much trouble and cost, the location of the defect was ascertained, but all attempts to recover the cable failed. In 1865 was commenced the laying of the second cable, and about half of it had been paid out when it broke. Operations were suspended until the following year (1866), when a stronger but lighter and more flexible cable was successfully laid, the distance between Trinity Bay and Valentia being 2,134 miles. In 1869 the French Atlantic line between Brest and St. Pierre, and thence to Duxbury, Mass., went into operation, and in the summer of 1875 the final splice of the Direct Cable Company's line was made. Since the days when the difficulties in the way of trans-Atlantic telegraphy appeared almost insuperable, wonderful strides have been made in the electric art, and great facilities have been introduced in the method of paying out the cable from the ship, so that what was formerly regarded as a vast experiment has now become a very practicable work.

MR. FIELD AND THE CABLE.

Mr. Field's energetic labor was pursued with a zeal which entailed heavy financial expenditure. Though a man of independent fortune when he began, he embarked in it so large a portion of his capital as nearly to make shipwreck of the whole. While in England, engaged in the expedition of 1857, a financial storm swept over this country, and his house suspended; but on his return he asked only for time, and paid all in full with interest. The stoppage, however, was a heavy blow, and, being followed by a fire in 1859 which burned his store to the ground, and by the panic of December, 1860, just before the breaking out of the war, he was finally obliged to compromise with his creditors. Thus released he devoted himself to the work of his life. The success of the Atlantic cable brought back a portion of his lost wealth, when his first care was to make good all losses to others. He addressed a letter to every creditor who suffered by the failure of his house in 1860, requesting him to send a statement of the amount compromised, added the interest for nearly six years, and as fast as presented returned a check in full. The whole amount is stated to have been $200,000.

CABLE OPERATORS.

These persons form a class by themselves, requiring a special education and special adaptability to the service. Their life is anything but a cheerful or social one, for they are usually located in out of the way places on the sea coast, where neighbors are few and far between, and scarcely of a character calculated to constitute an interesting and pleasant social circle.

When on duty they are closely occupied in watching

and translating the slender point of light whose vibrations convey to the eye with them, as sound does to the ear of the ordinary telegraph operator, the intelligence which it is necessary to communicate. When off duty their pleasures and recreations are few indeed, and taken altogether the occupation and it surroundings are not enticing to individuals of social and companionable proclivities.

It may be said on the other hand, however, that the labor required is not excessive, and is well paid. If there is a lack of opportunity for social enjoyment, there is also not much temptation to spend money, so that the position of cable operator is one in which there is an opportunity for financial accumulation. Most if not all of the cable operators on this side of the Atlantic came from England, and after a certain term of service are entitled to a three months' leave of absence to visit their native land, if they so desire. They receive from the company a liberal allowance to defray their expenses upon the trip.

ECONOMY IN SENDING MESSAGES.

The price per word being a consideration in transmitting messages over the Atlantic cable, the aim of merchants, news agencies, and others is to send as few words and convey as much information as possible. A great number of cipher codes are in use, composed generally of columns of words or figures answering to every possible emergency. The codes are kept profoundly secret, and to prevent the clerks and employes in the offices interpreting and divulging the message, a secret understanding often exists between the principals to read the cipher backward or forward half a dozen words. The following sample of a message presents the most unintelligible aspect to an outsider:

John Bolton & Co., Liverpool, to Preston, Banks & Co., New Orleans.—Kildare—Description—Sacred—Ecuador—Pot—Screamer—Shrimp—Betsy—Nameless—Bobby—Bellona—Obscure—Numantia—Rattletrap—Richard—Sackbut—Sally—Salmon—Penholder.

Such a queer combination of words might lead one to the conclusion that the cotton merchants were given to indulge in an eccentric species of wit peculiar to themselves; but the words have a stern significance that means "business." They form a cipher telegram of the most unrelenting "business aspect," even the diminutive "shrimp" bearing a grim message of special intelligence, and the very unsentimental Christian names answering to the names of various firms, who are wont to be addressed by much more respectful titles. It is necessary to take notice that the cotton bought by cable is still in this country or on the sea; in fact, it is often bought, re-sold, and re-bought again perhaps half a dozen times before it ever touches the shores of England. The translation of the telegram above given is as follows:

Kildare
Description
Sacred.

We have sold to Kingston & Co., Preston, 500 bales (of cotton) at 7¾ (per pound), good quality, color and staple. Terms laid down by steamer. Bills of lading to be sent through Messrs. Baring Brothers.

Ecuador Pot.	Buy for John Smith & Co., 200 bales at 8⅝, with fine, long, even staple; inferior bales will be rejected. Ship by steamer.
Screamer.	Execute this order if possible; it may lead to a large business.
Shrimp Betsy.	Do not insure for Brown & Co.; they will attend to their own.
Nameless Bobby.	Bush & Wilson are not satisfied with their lot; it is not up to the mark. Use more care. Take special pains to ship no bales showing sticks or sand.
Obscure Numantia	Your letter is not to hand; if important, cable particulars.
Rattletrap Richard	The Numantia is making a long voyage: fears are entertained for her safety.
Sackbut Sally	Is James Rochdale good? and to what amount? Sharp is speculating. Be careful with him.
Salmon Penholder	The Manchester market is excited and rising rapidly.

This cable telegram is a fair specimen of the kinds that are daily passing by hundreds over the Atlantic cable. The art of preparing these codes is one requiring considerable ingenuity.

HUMORS OF THE TELEGRAPH.

In the progress of this little book up to the point reached, occasion has been met with for the introduction of incidents and anecdotes which have served to lighten its pages and add to their interest. Notwithstanding this, however, our plan would not be complete in the absence of an entire chapter devoted to the humors of the telegraph, and giving a succession of well-authenticated accounts most mirthful and entertaining.

We can scarcely do better than to introduce, in the beginning, the ingenious Irishman who inquired of an operator: "Do you ever charge anybody for the address of a message?" "No." "And do ye charge for signing his name, sir?" continued the customer. "No." Well, then, will ye please send this? I just want me brother to know I am here," handing the following: "Cincinnati, Sept. 3d. To John M'Flynn—at New York—(signed) Patrick M'Flynn."

An old lady in a town of Massachusetts, refused the gift of a load of wood from a tree struck by lightning, through fear that some of the "fluid" might remain in the wood, and cause disaster to her kitchen stove. And during the summer of 1878, a Texas man declined to receive a dispatch from a yellow fever locality, lest he might catch the disease.

That was a witty man who, being detained by a

snow-blockade, penned a dispatch which ran thus: "My dear sir, I have every motive for visiting you, except a locomotive." So was the other who, under similar circumstances, telegraphed to his firm in New York: "I shall not be in the office to-day, as I have not got home yesterday yet."

Incongruous telegrams as to their subjects, are numberless, their reason, economy. For instance: "To ——. Nellie has fine girl. Sell my horse at price named."

Another, sent from a Western town to a gentleman of this city, read: "To——. Matilda died this morning. Send fifty dollars worth of cheap jewelry."

A message sent from Cincinnati to Milwaukee read: "Send Pauline here immediately; have a chance to get her married." And a Pennsylvania politician once telegraphed his father: "I have 2,000 majority—brother Sam died this morning."

From Albany, Oregon, we learn of a farmer down the country who had occassion to telegraph to that city to friends, notifying them of the death of his father. Being anxious to get the message through promptly, he rode on horseback past one telegraph office to another, twenty miles nearer Albany, to send the message, giving as his reason for the extra travel that the office was twenty miles nearer, and, of course, the message would go quicker than from the other twenty miles further away.

The following dispatch created no little amusement in the offices through which it passed. " Charlie and Julia met at S——'s yesterday, quarreled and parted for ever; met again this morning and parted to meet no more; met again this evening and were married."

There are evidences of a poetical turn of mind in this telegram, sent by a newly-married man while on his wedding tour, to a friend in Montreal: "Expect to-night a happy pair, bed and supper please prepare;" and of domestic bliss in the following, sent by a Wall street broker to his wife: "Send John. Also demijohn. Kiss Matty. Spank Arthur. Don't fret."

The husband of Harriet Prescott Spofford was in Boston when he learned that he had become a father by this dispatch, dated Newburyport: " Dear father, I came to town this morning at eleven o'clock, and when you are disengaged I shall be very happy to be introduced to you. Truly your affectionate son, Richard Spofford."

"Mamma," said a little girl, pointing to the telegraph wires, "how do they send messages by those bits of wires without tearing them to pieces?" "They send them in a fluid state, my dear," was the reply.

A good story is told of a country woman who received a dispatch later than she expected: "It must have been delayed on the road," said she. "I know the wires are busy to-day, for I heard them working as I came along."

"KILLING FAURE."

Much ado about a little arose from the meddling of an astute operator of Paris, who, upon receiving a dispatch of an unusual character for transmission, stared and inquired of the messenger by whom it was sent. The answer was: "By a gentleman living in the Rue la Fontaine." The operator requested the man to step into his office and take a seat. Meanwhile a gendarme was summoned and the message shown to him. It ran thus: "I have thought of a better and more expeditious mode of killing Faure," and was signed Mery. The agent started for M. Mery's residence; he was in bed, but was in the act of announcing to his co-laborer, M. Dulvile, with whom he was writing Don Carlos, for which Verdi composed the music, that he had thought of another mode of dispatching the Marquis of Rosa (which part was to be acted by Faure) than by a pistol-shot, as in Schiller's tragedy, and had telegraphed to him to that effect.

"ADDITIONAL WURRED."

"The top of the mornin' to yez, sur," remarked an Irishman, entering the Cincinnati office one morning.

"Good morning," replied the operator.

"Fhot do yez charge to sind a missige to Pittsburg?"

"Forty and three, sir."

"And fhot is the three fur, I dunno?"

"That is for the additional words, sir."

"Additional wurred! And who is he?"

"Why, for ten words you pay forty cents and for each additional word three cents."

"Oh, ho! ye spalpane! and that's your little game, is it? Yez wants me to pay yez forty cints, which yez will pocket, and thin sind the missige wid that three cints by mail, eh? Oh, no! I'll sind it by mail meself, and get tight on that same forty cints! Good day to yez."

And out he went, leaving the telegraphers to enjoy a hearty laugh.

A SKETCH FROM RUSSIA.

The last story is matched by one which reaches us from Russia, and is a faithful account of what took place in one of the Russian telegraph offices.

The door is opened by a stout merchant.

Merchant—Hollo, there! Is it here that you send telegrams?

Operator—We can dispatch a telegram for you, sir, if you wish it. Will you be so good as to write down the message that you want to send?

The merchant took a sheet of paper, sat down with an air of stern satisfaction, and wrote:

"To my son, Vasili Petrovitch Bogatoff, at Moscow: Vasia, you infernal dog! You fool, you pig, you villain, you brigand, you pickpocket, you unbaptized son of a gun! What the devil do you mean by rousing me up in the middle of the night with that cursed letter of yours, begging for money, as usual? Not a kopeck shall you have from me, and you may go and hang yourself!"

Operator (mildly but firmly)—Excuse me, sir, it is quite impossible for us to send such a message as that.

Merchant—How? Not send it? What do you mean? If I were to put that in a letter and mail it, it would go, and why shouldn't it go in a telegram? Besides (with an air of unanswerable logic) he *is* a pig! Come, you must send it—you know it's your duty.

Operator (with exasperating politeness)—Quite out of the question, sir, I assure you. Our rules are very strict, and we never depart from them.

Merchant (furiously)—So much the worse for you, then. I'll write a letter twice as bad as that message, and send it off by the first mail—and then we'll see. That for you and your telegram! They're not worth a kopeck.

Exit triumphantly.

A SATCHEL BY TELEGRAPH.

The subject of our story was a German somewhat intoxicated, who boarded the Hudson River train at Kinderhook. He threw his satchel down in a corner of the car, took a seat, and was soon in the arms of Morpheus. On awaking he alleged that he had left his baggage at Kinderhook, and asked the boy employed on the train what he should do to recover it. The latter, who had seen the German place his satchel in the corner, replied: "You give me thirty cents and I'll telegraph to Kinderhook to have the depot master forward it by telegraph to Greenbush. It will reach there before we do." The German paid the money, gave a minute description of the missing property, and the boy departed, taking the satchel into another car. On reaching Greenbush the boy returned with the bag, and, placing it in the German's hands, said: "There's the first satchel I ever see come by telegraph." "Ah," replied the German, "dot delegraff is vun great dings; here, dake another quarter, mein boy." And the boy did.

A MEDDLING KING SNUBBED.

King John of Saxony was prone to dropping in upon officials when they least expected him. One day he appeared at the telegraph office of a small station. The operator apprised his colleague at the next station of the unwelcome visit, and before an acknowledgement of the warning came, was called upon to enlighten the inquiring monarch respecting the business of his office. Presently a message came along the wires, and his majesty desired to be acquainted with its purport. He was told it was unimportant, but was not to be put off, and insisted upon the message being repeated to him; so the stammering operator had no choice but to regale the royal ears with the German equivalent for "The king pokes his nose into everything."

A VERY PROPER OLD LADY.

A droll mistake was made by an imaginative old dame who, having permitted a telegraph pole to be placed in front of her house, waited on the chief of the telegraph company concerned to complain that she could get no sleep at night, being kept awake by the noise made by the messages passing over her head.

"I don't think, sir," said she, "you can be aware of all that's said along them wires. There's a deal that hadn't ought to be. I can assure you, sir, that very much that's said there, that I have to lie and listen to, is such as no decent woman ought to hear; and I hope you will put a stop to it."

The amused gentleman was hardly able to meet the accusation with due gravity; but he did contrive to keep his countenance while he informed the old lady that the young men who had hitherto worked the wires were under notice of dismissal; and that in future only young women of great respectability would be employed, so there would be no danger of her propriety being shocked any longer.

LITTLE "JOHNNY RUSSELL."

One evening at a time when Lord John Russell, known in English public-house political disputations by the disrespectful name "Johnny Russell," was in attendance at Queen Victoria's castle of Balmoral, in the north of Scotland, a little old man, buried in a great coat, handed a telegram, addressed to one of the ministers in London, to the telegraph operator at one of the stations on the Deeside railway. The operator, after glancing at the message, threw it contemptuously back with:

"Put your name to it. It's a pity your master does not know how to send a telegram."

The name was added.

"Why, you can't write!" exclaimed the operator, after vainly trying to make something of the signature. "What's your name?"

"My name," said the messenger—"my name is John Russell."

That operator was transferred to another office before many days passed.

PETER'S TELEGRAM.

A message had been received for Peter from a former sweetheart, Margaret Flagarty, inviting him to spend the day with her. Of course the telegram was duly sent to his address. That evening a forlorn-looking object entered the office, and going to the operator, said: "Please, sur, I want to send a message." "Well, here is the paper, write it down." "Indeed, sur, I can't write." The operator, who was a brisk little man, said: "Come to the desk, then, and tell me what you want to send." He came slowly, and gave the address of Margaret Flagarty, etc., then, in a deep, sepulchral tone, hitching nearer the instrument, he added: "I am married, and to my sorrow!" If the wires didn't laugh the operators did, as the message sped swiftly from station to station. No two-volumed novel, with connubial miseries long drawn out, could have portrayed more heart-rending grief than Peter's telegram.

HE COULDN'T BE FOOLED.

"Would you mind readin' this for me, sir? I can't read myself." It was a snow-shoveler on Walnut street, Louisville, that spoke, as he handed over an envelope, inclosing a telegram, which read: "Nashville, January 9, 1879. I shall arrive at Louisville by the three o'clock train this evening. Jerry A. Taft." "Will you read it again, sir?" asked the snow-shoveler. It was read again. "You say it's signed Jerry A. Taft?" "That is the name." "Please read it once more." His request was complied with. "It goes right straight along—just them 'ere words, without any hitchin' or stumblin'?" "Just that way." "It can't be Jerry, then; it can't be Jerry," he mused; "Jerry couldn't say that many words without stutterin' all to pieces, to save his life. Some fellow's tryin' to fool me, but I'm too smart for him, I am."

WRITES LIKE A MAN.

A family in the country were electrified by the receipt of a telegraphic dispatch from a daughter, who was teaching in a distant city. The telegram was passed around and duly admired. The dashing boldness of the chirography came in for its share of the praise. The old lady shook her head with an air of gratified pride, as she ejaculated, slowly:

"Anna Maria allers did write like a man; she's been takin' writin' lessons; this here beats her last letter all holler!"

A LITTLE STORY FROM MAINE.

A man went into one of the offices in Bangor with a dispatch, which he insisted upon having sent off immediately. The operator accommodated him, and then hung the dispatch on a hook. The man hung around some time, evidently unsatisfied; at last his patience was exhausted, and he belched out: "Ain't you going to send that dispatch?" The operator politely informed him that he had sent it. "No yer ain't," replied the indignant man; "there it is now on the hook."

HOLLOW AND HELLO.

A genuine "pahdee," quite aged, living some miles out of town, went into the office at Augusta one day to sell some "praties," and seeing the instruments, battery, etc., wondered if that was the "tiligraft." After gazing steadily for several minutes, he said he had always wanted to ask one question; and this was it: "Is the wire hollow on the outside or on the inside?" Some one recently inquired of the manager of a telephone exchange whether telephone wire wasn't hollow. "No," gruffly replied the manager, "it's 'hello.'"

FOOLING THE SAVAGES.

The ingenious French have contrived a novel way to impress the barbaric mind. M. de Brazza, who has charge of the expedition to Senegal, carries an electric battery in his pocket, communicating with two rings on his hand and with other apparatus scattered about his person. When he shakes hands with a savage chief, that chief will be very much astonished, for an electric shock will run up his arm and he will see lightning playing about the head of his visitor. Naturally he will think he is being interviewed by his satanic majesty, and will be ready to consent to anything in order to get away.

"ONNATERAL FIXINS."

An old lady living on one of the telegraph lines leading from Louisville, in the early days of telegraphy, observed some workmen digging a hole near her door, she inquired what it was for. "To put a post in for the telegraph," was the answer. Wild with fury and affright, she seized her bonnet and ran off to her next neighbor with the news. "What do you think?" she exclaimed in breathless haste; "they're setting up that paragraph right agin my door; and now I reckon a body can't spank a child, or scold a hand, or chat with a neighbor, but that plaguy thing'll be blabbing it all over creation. I won't stand it. I'll move right away where there ain't none of them onnateral fixins!"

CHICAGO AND ——.

During the time when the Atlantic and Pacific Telegraph Company had established a uniform rate of twenty-five cents between any two offices east of the Mississippi River, a Chicago man, residing in the suburbs, having to telegraph home from a distant Wisconsin town, asked the diminutive and apparently unsophisticated operator in charge of the only telegraph office in the place—a Western Union one—what the tariff would be, and upon being told that one insatiate dollar would suffice, burst out: "Dollar be blowed! We can telegraph to h—l in Chicago for a quarter!" "Oh, yes," calmly answered the cunning knight of copperas and brass, "but that ain't outside the city limits!"

A WITTY ILLUSTRATION.

Writing of the difficulty English engineers experienced in making educated Persians understand the working of the electric telegraph, Mr. Mounsey says: "Much of the time of one of our officers was occupied during several weeks in attempting to enlighten the mind of a provincial governor, who had got it into his head that the wires were hollow tubes, and that messages were transmitted through them, as in the pneumatic post. In vain was the whole apparatus shown to his highness; in vain even all its parts explained and re-explained—he stuck to his idea; and it was only by the suggestion of the following simile that he was at last induced to relinquish it, and declare himself satisfied:

"'Imagine,' said the officer, 'a dog whose tail is here at Teheran, and his muzzle in London; tread on his tail here, and he will bark there.'"

OFFICE LOAFERS ELECTRIFIED.

Newspaper editors especially will be thankful for a description of the manner in which certain telegraph operators of Sacramento, Cal., rid themselves of loafers. A box running the full length of the front of the office on the outside had furnished a tempting seat for the habitues. This was covered with zinc, which had been connected with the batteries that were contained in the box. A person sitting upon the box without touching his hands thereto did not feel the electricity, but if his hands dropped on the box, or he put them thereon to assist him in rising, he received such a sudden and astonishing shock as sent him an unbelievable number of feet toward the lofty roof and the adjacent river. Any good day a person might see some of these unfortunates, unexpectedly struck with this domesticated lightning, describing a fifty feet parabola in the air.

SHOCKING THE NEGROES.

At one of the stations on the Kentucky Central Railroad, a couple of negroes cut down a tree across the telegraph wire and broke it. The operator came out at once, determined on revenge. He quietly took his seat and ordered the negroes to bring the two ends of the wire together and mend it. Each seized end and end, but the moment they came in contact there was a sharp electric shock, and they let go. It was raining, and the battery was strong. However, the negroes didn't know where the shock came from, and tried it again. By this time they were so wet that the current would pass if the clothes of one but touched the other. Frightened and bewildered they brought the wires together again and again; each time, to their great astonishment, an electric shock convulsed them. And when the train started there sat that operator under the shelter of the depot still egging the negroes to fresh efforts.

BLINDFOLDING THE "MASHEEN."

Mrs. Moore, desiring at times to indulge in a little domestic telegraphy, had a wire run from the basement of her domicile to the second story sitting room thereof, and equipped with a pair of learners' instruments. By the help of a telegraphic friend she and her husband soon learned to communicate deftly with each other, sending down instructions to the servants and superintending household matters generally without the inconvenience of traveling too frequently up and down two flights of stairs. Bridget and Mary, of the lower regions, had watched this mysterious operation with considerable interest, and, as the event proved, had settled upon a theory of their own as to the *modus operandi* of the concern—at all events they evidently considered that it was not altogether a safe thing to have in the room under certain contingencies.

One evening Patrick and Michael had paid a visit to the aforesaid handmaidens, and the quartet had remained in close conference with closed doors until a late hour. The next morning Mrs. Moore discovered the telegraph instrument carefully covered over with a cloth, and nicely tucked in around the edges! At first she was naturally astonished at such unprecedented care-taking, but when the truth flashed upon her that the unoffending instrument had merely been blindfolded, so that it couldn't see what was going on and report it to the "missus" up stairs, she laughed till she well nigh went into convulsions. So do her friends when she tells them the story.

A CRAMMER.

The buffaloes found in the telegraph poles of the overland line a new source of delight on the treeless prairie—the novelty of having something to scratch against. But it was expensive scratching for the telegraph company; and there, indeed, was the rub, for the bisons shook down miles of wire daily. A bright idea struck somebody, to send to St. Louis and Chicago for all the brad-awls that could be purchased, and these were driven into the poles, with a view to wound the animals and check their rubbing propensity. Never was a greater mistake. The buffaloes were delighted. For the first time they came to the scratch sure of a sensation in their thick hides that thrilled them from horn to tail. They would go fifteen miles to find a brad-awl. They fought huge battles around the poles containing them, and the victor would proudly climb the mountainous heap of rump and hump of the fallen, and scratch himself into bliss until the brad-awl broke or the pole came down. There has been no demand for brad-awls from the Kansas region since the first invoice.

Right here we must shut down on funny stories. 'Tis time to retire.

TELEGRAPHIC "BULLS."

This is a fruitful section, probably to many readers the most interesting of all, if not the most useful. We must, however, keep it within reasonable bounds, culling from the best "bulls" which have come within the writer's knowledge, and telling these as concisely as possible, so as to include a goodly number.

In so doing we find it the most convenient way to divide this section into two parts—telegraphic "bulls" by operators and by the public.

"BULLS" BY OPERATORS.

"Bulls" are not all of a funny character. How big with fate to the last French empire was the telegraphic blunder which caused the defeat of Marshal McMahon, in the summer of 1870! Failly had been telegraphed to move on Limbach; but the dispatch, as received by him, read "Kausbach," and he acted accordingly, by which move the plan of the campaign was fatally disarranged.

Perhaps Fritz, in the following story, taken from the history of the Titanic struggle in the first year of the present decade, deserved, for his mercenary view of marriage, all the inconveniences and the disappointment which a telegraphic "bull" caused him.

A young German lieutenant, wounded in the Franco-German war, went for his health's sake to a quiet village in Vaud, where he found a sweetheart. By the time he had regained his health the pair were engaged; then came a sudden order to report himself at Berlin, an order he, of course, obeyed. At first his disconsolate Marie was comforted by frequent letters full of protestations of love and constancy; but as time wore on the lieutenant plied his pen less often and moderated its outpourings. At last he suffered six weeks to go without a word. He was expecting a reproachful reminder, when a telegram arrived from the faithful girl, which may be thus translated: "Dear Fritz,—I have just received a letter informing me that my uncle, a millionaire in the East Indies, is dead, and that I am his sole heiress." Fritz felt his love revive as he read. He applied for leave of absence, and was soon exchanging greetings with the Swiss maiden. Though the coming of her lover filled her heart with joy, she could not refrain from gently upbraiding him for his silence. "Don't let us speak of it, dearest," replied he. "There is no longer any obstacle to our union. The unexpected good fortune which Providence has sent us has removed the objections of my parents; a fortune so colossal——" "Fritz," interrupted Marie, "do not make fun of me." For answer the lieutenant drew her telegram out of his pocket, and showed her the words: "My uncle, a millionaire in the East Indies, is dead." The poor girl, dropping his hand, said, "Dear Fritz, I wrote: 'My uncle, a *missionnaire.*' He has left me all he had, which is just a hundred and ninety-six francs." Fritz went back to Berlin freed from his engagement.

A writer on the other side of the Atlantic charges operators with having amazed a husband on his travels by informing him that he was the father of a *dolphin;* with having *extinguished* (distinguished) a man in Paris with an enormous red cockade; made Italy pregnant with *a lamb* (alarm); sent a man a train filled with *penny shovels* (perishable goods); told one man that his *onions* (opinions) were not wanted; made travelers inform their employers that they could not leave London without their *cabbage* (luggage); asserted that *sugar cans* (canes) grew in Jamaica; that *seraphs* (serfs) were emancipated in Russia; that the Emperor of Austria gave the ambassadors a *spree* (soirée); made Captain Smith, of Her Majesty's 33d, indignant by addressing him as Captain Smith, of Her Majesty's dirty 3d; amazed a distinguished poet by consigning to him a cargo of codfish and salt pork, and amused a distinguished clergyman by asking him his lowest offer for steam coals; and nearly got a merchant into the "black list" by saying that he was *nowhere* (now here.)

Considering the many millions of messages sent and received every year—some operators in the larger offices handling as many as from three to five hundred a day—and the fearful and wonderful penmanship in which many of them are disguised by the senders, the wonder is not so much that mistakes occasionally happen, as that they do not occur far oftener, especially as the telegraphic symbols for many different letters and words are so nearly alike.

The most frequent cause of error on the part of operators is the running of two or more words together, on the one hand, or the unnecessary dividing of a word, on the other. For instance, the words "colored man" have been transformed into "Col. Ordman;" "Addie Pratt" into "Addie P. Rat," and the signature "Theodore Rose" into "the odor of roses."

"Subpœna witnesses and compel attendance" was made to read "Subpœna witnesses and compel Allan to dance."

"Your son is dead. Be at depot. Will arrive to-night," was changed in transmission to "Your son is dead beat. The depot will arrive to-night."

A gentleman was once considerably surprised to receive the following: "Do not hang about the hotel. Will write." The original message read: "Do nothing about the hotel. Will write."

A newspaper dispatch published some years ago gave an account of the doings of a number of troops under the leadership of A. N. Cushman. As printed in the papers, however, it stated that the troops were led by "an Irishman."

A story is told of a Kalamazoo, Michigan, judge who went to a neighboring town on business, and telegraphed back to his wife: "Have found Garland. Won't be home for a week." When received, the message read: "Have found girl, and won't be home for a week," which doubtless made an explanation necessary when he did get back.

The following dispatch was recently sent by a lady

to her reverend husband, who was off on a visit: "Come home and marry M. E. Stuart Thursday morning." The worthy divine received the message in this shape, which considerably startled him: "Come home and marry me. Start Thursday morning."

To properly appreciate many good "bulls" it is necessary that one be acquainted with the Morse telegraphic alphabet. It is believed, however, that the following will be found interesting even to those who do not know anything of the business:

There are two hotels in London much frequented by gentlemen of the bar. One is Thavies' Inn, and the other Sergeant's Inn. In a telegram addressed to a disciple of Blackstone at the former house the name of the hotel was rendered Thieves' Inn, and, curiously enough, about the same time another telegram called the other house Serpent's Inn.

A merchant in Boston recently received the following dispatch:

"CHICAGO, July 24.

"Jennie is good—now six dogs regularly."

His surprise was great. What Jennie was good for he could not imagine, and six dogs regularly was incomprehensible, unless it referred to diet, and then it was monstrous and astounding. After some conjecture he telegraphed for an explanation, and was relieved by the following correction:

"CHICAGO, July 24.

"Time is good—now six days regularly."

The subject in question was the time occupied in

shipment of goods to the West. Jennie was an irrelevant female introduced by the operator; and as for the dogs, they were a pure invention.

An English lord, as proud and fond as a man should be of his beautiful young wife, was just about rising to speak in a debate in the House of Commons, in London, when a telegram was put into his hands. He read it, left the House, jumped into a cab, drove to Charing Cross, and took the train to Dover. Next day he returned home, rushed into his wife's room, and, finding her there, upbraided the astonished lady in no measured terms. She protested her ignorance of having done anything to offend him. "Then what did you mean by your telegram?" he asked. " Mean ? What I said, of course. What are you talking about?" "Read it for yourself," said he. She read: "I flee with Mr. —— to Dover straight. Pray for me." For the moment words would not come; then, after a merry fit of laughter, the suspected wife quietly remarked: "Oh, those dreadful telegraph people! No wonder you are out of your mind, dear. I telegraphed simply: "I tea with Mrs. —— in Dover Street. Stay for me.'"

Sometimes operators are called upon to pay for losses that may be occasioned by mistakes made in messages received by them, as in the following:

They called him "Towser," and he was making frantic efforts to get up a reputation for never breaking. One day as he was passing a certain desk he heard a call, and gracefully vaulted upon a high office stool to answer it. This is how he copied the message:

"To John Brown, wholesale druggist.—Please send per express one barrel bottled ale immediately.

Seaton Bros."

Bottled ale was not in Mr. Brown's line of business, but Seaton Brothers were old customers of his, and so, willing to oblige them, he procured the ale and forwarded it without delay. Next day, in return for his kindness, they sent him the following message:

"To John Brown.—What do you mean by sending us ale? We refuse it. Hurry up our oil.

Seaton Bros."

Surprised and indignant at their apparent ingratitude he hastened to the office and wrathfully exclaimed: "What in the thunder is the meaning of this? There's been a lovely blunder made somewhere! Get that message repeated quick!"

So they got it repeated, and it turned out that it was a barrel of boiled oil Seaton Brothers wanted, instead of bottled ale. When this was explained to Mr. Brown—they broke it to him as gently as possible—he did not fly into a rage with the long-suffering manager, as they expected him to do. He merely remarked: "That operator must be pretty fond of ale when he takes to dragging it into messages so promiscuous like. However," he added, grimly, "he shall have plenty of it for once, for he's got to take that barrel and pay for it, too. Yes, sir, pay for it!" he repeated, with savage emphasis.

Another instance of a little different nature: One evening the proprietor of the railroad eating-house at

Summit, California, received the following dispatch: "Have 100 gallons coffee for my men on arrival of No. 1. (Signed) Lieutenant Morgan, Commanding detachment Co. B."

The operator promptly delivered the message. A happy smile overspread the landlord's countenance, for he had had government contracts before. He grasped the dinner gong, and never before did that gong give forth sounds so loud and long. It quickly summoned to his side half a score of cooks, waiters and maids; the order of the night was read, and each assigned to a post of duty. All was bustle and confusion. Being only an eating place for train men and passengers, the stock of tinware and cooking utensils was not very extensive. The landlord skirmished around the premises for tinware, and in lieu of coffee pots, etc., clothes boilers, dish pans, milk pans, dippers, and even oyster cans were filled with water and ground coffee, and placed upon any available spot where heat could be transmitted to their contents. Quantity not quality being desired, even the operator utilized the wash basins, and made three gallons over his office fire. What hurrying, shouting and swearing! Everybody got soaked with coffee; everything that would hold fluid contained coffee; even the china pitchers and wash basins in the rooms fitted up for the accommodation of guests had to be used.

Fifteen minutes before the train was due the landlord found that he had the required quantity all made, and was proud of his success. The train arrived.

Lieutenant Morgan, accompanied by two men, each carrying a five gallon can, entered the hotel. The cans were quickly filled, and the men departed. "Bring on your other cans," shouted the landlord. "What other cans?" asked the lieutenant. "To hold this coffee you ordered," replied the landlord. "I ordered?" and the officer gazed about him in astonishment at the array of cans, crockery and waiters. "Yes," shouted the landlord, drawing forth his message and exhibiting it. "You ordered *one hundred* gallons of coffee." "I ordered but *ten* gallons, and here's your money for it," replied the officer, throwing down a five dollar greenback. "All aboard," shouted the conductor, and the lieutenant rushed from the room. The landlord was now frantic; he quickly followed the officer out, but the train had started, and in a few moments was thundering down the mountain side a mile away.

Then the landlord swore, and made for the telegraph office. A very emphatic, if not elegant, salutation fell on the operator's ears. He was astounded. He immediately called up the office from which he had received the message, and had it repeated. Sure enough it read ten. The upshot was that the operator had to pay eighteen dollars for the ninety gallons of coffee.

Occasionally, however, mistakes of this kind turn out to the advantage of the customer, and no complaints are made. A merchant once telegraphed to a wholesale produce firm in New York to buy him a quantity of cheese. The original message said a hundred, but as delivered it read a thousand. Knowing the man to be perfectly responsible, the firm purchased for and sent him all the cheese it could get. The merchant thought that so much cheese would ruin him; but it so happened that the unusual demand had the effect of increasing the price to such an extent that he was able to sell it again at an almost fabulous profit.

In the summer of 1864 a telegraphic order was sent from Washington by General McCallum, superintendent military railroads, to Major Wentz at Binghamton, N.Y., to forward one hundred and fifty railroad men to Washington at once. The dispatch, when it reached its destination, read "fifteen hundred men." Such a demand was considered extraordinary, but in those days of "military necessity" strange things were always expected, and the men were soon collected and on their way South, wondering into what part of Dixie they were to clear a way for Uncle Sam's iron horses. But the surprise of the superintendent was still greater when they arrived, and a search was immediately instituted for the operator who made the mistake. As it cost about thirteen thousand dollars to transport the men to Washington, and the expense of keeping them there was not less than two thousand dollars a day, it seemed likely to prove a serious affair for somebody. It was ascertained that the error occurred in transmission between New York and Binghamton; but before the investigation was concluded, an order came from General Sherman, then at Dalton,

Georgia, to send him one thousand railroad men *immediately*, and so the blunder resulted in good to the government, and the telegraph was saved from censure.

"BULLS" BY THE PUBLIC.

All telegraphic "bulls" should not be fathered upon companies and their operators. The public are responsible for a large share of them. One principal cause of this is the miserable manuscripts furnished operators by customers. The following is a case in point:

An eminent divine was to deliver a lecture in a neighboring city, and wishing to telegraph his subject ahead for advertisement, hastily penned a dispatch, handing it to a boy to deliver at the telegraph office, he himself leaving town. The operator, after sinking a shaft of close scrutiny into the Chinese-like hieroglyphics of the message, seemed suddenly to strike a vein of intelligence, and the message went quickly on its way, the subject of the lecture being duly announced in the next morning's paper as "Our Constitutions, and Fresh Halibut." The sender of the message, who had come to lecture upon "Some Considerations on the Force of Habit," says if anybody will start a petition to suppress all telegraph companies, he will be the first to sign it.

Correspondents of the press, when they use the telegraph, are in the habit, for economical reasons, of dispensing with articles, prepositions and conjunctions, while punctuation is perforce out of the question; and the "bulls" arising from this cause cannot fairly be blamed on operators.

Of such was that occasioned by a message sent from England to the editor of the Java *Bode*, which read: "Proposed to Brand Speaker," meaning that Mr. Brand had been nominated Speaker of the House of Commons. Printed as above, the meaning conveyed to the readers of the journal was that it was proposed to brand the Speaker of the august body indicated.

"Bulls" in original messages might be given that are fully as amusing as any made by operators. For example: "My barn burned up last night, October 22. I want you to come and see it." Or the following, sent from Kingston, N. Y.: "To J. W. B., Honesdale, Pa. Your horse died this morning after writing you a letter."

To show how difficult it is to make out some of the words in messages, and how easily mistakes may arise, we give the manner of spelling a number of common words, as found in the dispatches of many patrons of the telegraph.

"Comerchel Worf," "Comerciol Warf," "Centrel Deapot," "Junktion," "Jursy Citty," "Nigra Falls," "Porkepsee," "Moris Weight Peches," "Pees," "Redash," "Turnups," "Cllamns," "Eells," "Ells," "Hadic," "Macril," "Ancer," "Ansewer," "Amediately," "Ameaditley," "Imegitlay," "Busnes," "Cittifacat," "Carridge," "Delade," "Dolors," "Evrey," "Garrentee," "Pararie," "Possable," "Pituculars," "Resons," "Speshall," "Spetial," "Seckend," "Two-day," "Two-knight," "John ded will bey berred tomorrough," "I will gow met me at depow." An erudite Assemblyman says his "Comity is tring to do so."

Sometimes most entertaining "bulls" have arisen from sheer carelessness on the part of senders, as in the following instances:

A merchant away from home received a telegram announcing that his wife had been safely delivered of a little girl. Simultaneously a message came from his partner stating that a draft had been presented to the firm with a doubtful signature, and inquiring if he knew anything about it. He at once replied to both messages, but somehow misdirected them. The amazement of the wife might be conceived when she was informed: "I know nothing about it: it's a swindle;" and of the partner when he received hearty congratulations upon his safe delivery.

An enterprising fish dealer in an eastern city indited a fish order to "Paine Brothers, Eastport, Maine," but his clerk inadvertently made the message read "Paine Bros., New York," a firm priding itself upon filling every order. Consequently the fish was sent from New York, arriving fresh and nice, but with a "C. O. D." attached, involving a bill of expense which the enterprising fish dealer declared the telegraph company should pay, or he would bankrupt the whole concern, if it took every dollar he was worth in the world.

Operators could tell of meannesses on the part of the public, occasioning errors, wrongful blame, and sometimes more serious consequences, almost incredible in their degree of contemptibleness.

"What means it," says a faithful manager of an

office, "that Mr. —— should come to us and demand that we refund the money he paid on that message you sent him? He says you paid for the message when you sent it." "I'll tell you how it was," says the patron, confidentially; "I ought to have paid for it—didn't want to look mean, you know, so I gave him to understand, in a roundabout way, that I did pay. Better be on the telegraph company than on me, you know; so you keep mum, it's all right."

Another example is that of a careless fellow who neglected till the last moment to answer an important telegram, and then, to cover his delinquency, replied by telegraph: "Did not receive your message till too late; train had left." "You see," he explained to a person accompanying him to the office, "I don't want to go, and there's no other way for me to get out of it." His friend, who had waited all day for the reply, vows eternal vengeance on the telegraph generally, and especially to that "contemptible apology for a manager who would let an important message lie around all day before delivering it!"

A correspondent of *The Operator*, Mr. D. C. Shaw, relates effectively the sad results of an error on the part of the sender of a message, with which account we must conclude this chapter.

"I was once at a small railway station," writes he, "and saw, on his way to the village hotel, a distinguished passenger whose leg had just been crushed by a moving train. All that skill and friendly services could do were instantly in operation. Sympathising and zealous young persons, at the sufferer's request, flew to the telegraph office to summon the wife. Full of excitement they write a message. A letter is omitted from the address, a single letter. The message is rushed to its destination, but—and you know the sequel. An hour passes; then comes an office message, 'Give better address.' The same name is given, with the same fatal omission, but they add the words, 'Care of Messrs. ———, No. —.' Another pause. Then another office message: 'Messrs. ——— have closed office and gone home to ——— (a suburban town); shall we deliver by special messenger?' Meantime a train arrives—the train upon which the wife should have come. The sufferer rouses himself expectantly. How hard it is to tell him she hasn't come! Then he fails rapidly, and they fear the result. Meantime the message is delivered; the wife is coming, but he is unconscious. And oh, the anathemas that pour in upon the telegraph and all connected with it! As the facts are known to but few outside of the circle especially concerned, the circumstances are misconstrued and exaggerated, and the poor operator, who would willingly have run with the message through all the hours of the day and night to insure its safe delivery, is branded as 'cruel,' 'barbarous,' and remains thereafter under a certain weight of ignominy through many unjust accusations."

LIGHTNING FREAKS AND TRAGEDIES.

As has often been said of fire and water, the electric fluid is an excellent servant, but a very bad master.

DEATHS FROM LIGHTNING.

Many persons are killed by it every year, probably more than is popularly supposed. According to some recently published statistics, more than ten thousand people have been smitten by the electric fluid within the past thirty years, of whom twenty-two hundred and fifty-two were killed outright. Of the eight hundred and eighty killed within the last ten years, only two hundred and forty-three were females.

The would-be wit of newspaper scribblers has been exercised upon this difference, the reason of which is clear when it is considered that men are exposed to accident far more than women are, because they spend less time at home, being abroad in the pursuit of business or in labor.

It would be well to bear in mind that persons struck by lightning should not be given up as dead for at least three hours. During the first two hours they should be drenched freely with cold water. If this treatment fails to restore animation, salt should be added to the water, and the drenching continued another hour.

EFFECTS OF LIGHTNING IN DIFFERENT COUNTRIES.

A difference in the effects of lightning in various countries has been remarked. It is said to be more dangerous in England than here. Why, so far as our knowledge extends, nobody appears to offer a reason. In France the mortality from lightning is twenty-seven a year on the average, of about two hundred and fifty struck. The low lying departments have fewer cases than hilly districts. Eighty were wounded and nine killed in one thunderstorm at Chateauneufles-Moutiers in 1861; and within one week, when the air was highly charged with electricity, thirty-three fearful flashes of lightning were observed, each bringing death to some victims. Nine deaths a year from lightning are reported from Switzerland, and but three from Belgium, a more populous country, which confirms the alleged greater frequency of casualties from lightning in hilly or mountainous districts—a distinction, however, which cannot be applied to England.

AN OLD NOTION EXPLODED.

The popular belief that when one gets into a feather bed he is safe from the ravages of lightning was rudely shocked by what occurred not long ago in a country store in Virginia which was struck by lightning. The fluid made a large hole in the roof, and passed through a feather bed, the recognized non-conductor. Happily no one was lying upon it.

A TRIPLE TRAGEDY.

During a severe lightning and thunder storm at Newberne, N. C., in the summer of 1878, three young persons, Isaac Richardson, aged twenty, Eliza Collins, twenty, and Laura Williams, nineteen, were struck by a heavy discharge of electricity and instantly killed. Richardson was escorting the two girls, one on each arm, from church to their homes, and as they neared Queen Street, a gentleman, who was but a few feet behind, saw them fall as a lightning flash struck them. The coroner found the lifeless bodies lying side by side, with arms still locked. At the time of the accident they were walking under a steel-handled umbrella, which was found lying upon the ground near the bodies, the cover partially burned, and which, undoubtedly, was what attracted the electric discharge.

SINGULAR FREAKS OF THE LIGHTNING.

A gentleman, while walking the streets at Des Moines, Iowa, during a thunder storm recently, had one of his eyes completely destroyed by lightning, without receiving other injury.

A queer freak of the lightning is reported from Rockville, Conn. It entered at the door of one of the stores in a livid flash, which actually lit an oil lamp, and left it burning, without leaving any other visible marks of its passage.

While a body of two hundred men were drilling at West Point, on one occasion, a black cloud, very low down, suddenly discharged itself of its electricity, seemingly through the attraction of the two hundred bright gun barrels, and the shock distributed itself throughout the corps. Several of the men were stunned, and a large proportion of the guns were knocked out of their owners' hands.

Lightning at Madison, Wisconsin, during the winter struck into the lake, and hurled masses of ice two feet thick hundreds of feet through the air.

Lightning recently struck a wine cellar in France, and converted a large quantity of bad wine into excellent brandy, a change appreciated by the owner.

During a heavy thunder shower at Mechanic Falls, Maine, in the summer of 1880, a boy was sitting at the foot of a Balm of Gilead tree which was struck by lightning. The tree was splintered, but the boy was apparently uninjured. Soon after the accident he was seized with nausea, and on a physician removing the little fellow's clothing there was found upon his stomach and chest an imprint resembling the trunk of the tree, its branches and buds as perfect as could be drawn by the hands of a skilled artist.

A thunderbolt which came down at Milton, Conn., and paid particular attention to the house of a Mr. Brown, deserves record for its singular and vigorous behavior. It began by demolishing the lightning rod in the most sarcastic and scornful manner. It then entered a second story room of the house, cut a hole six feet square in the floor, demolished the stove, and broke every pane of glass in the window, after which it mildly entered the dining room and ripped up the floor there. It made minced meat, so to speak, of the wash room, and left the house without any underpinning to speak of. Then it paid its respects to the barn, went back to the house, and violated the sanctity of a servant maid's room. The poor girl was just innocently adjusting her hair in the looking-glass when she was thrown violently backward on the bed by the furious thunderbolt, and she says she will never be vain again. In a neighboring house the frisky element " scattered a quantity of soft soap, and tore one rivet from a frying-pan."

LIGHTNING IN TELEGRAPH OFFICES.

Although a telegraph office is one of the best places to take refuge in during a thunderstorm, lightning sometimes follows the wires into the offices. During severe storms telegraph offices are generally " cut out." The switch-board, through which the wires pass before reaching the instruments, is provided with what are called "lightning arresters," so that but little damage can be done. Two cases of death by lightning in telegraph offices are on record, both of which occurred in the summer of 1876. One was that of a young woman in Nevada, and the other a Miss Clapp, manager of an office in Massachusetts. The latter had the instruments "cut out," but the lightning came in through the open window, there being a strong draught through the office, which, of course, should not be permitted during a thunder storm.

SHARP PRACTICE BY TELEGRAPH.

In common, we suppose, with every appliance of modern civilization, the telegraph is abused by the Ishmaelites in the community, who prefer rather to plunder than to work honestly. Numerous examples of this, some of them indicating remarkable ingenuity on the part of the swindler, are published from time to time as they occur.

One of these devices, unearthed at St. Louis, consists in bringing two telegraphic dispatches and a messenger's book to a wealthy man for his signature, the page of the book being so cut and underlaid with a blank check that the signing of the name twice would give the clever operator a check both signed and indorsed. One business man narrowly escaped the trap, which failed for lack of a little forethought, as the paper beneath, not being securely fastened, slipped enough to attract attention as the name was being signed the second time. This small circumstance defeated the plan, and saved the discoverer a big deficit in his bank account.

OBLIGE THE GENERAL.

On one occasion, when General McClellan was in Europe, many prominent New Yorkers received, as they supposed from him, cablegrams to the effect that, having purchased a horse, for instance, from say John Smith, for $420, it would be considered a particular favor if the person addressed would pay Mr. Smith the amount, which would be made right on the general's return. This dispatch was usually delivered in the forenoon, while Mr. Smith made his appearance in the afternoon with the bill and presented a telegram purporting to be signed by General McClellan, requesting him to call at that address for the amount. The money in nearly every case was paid, and in this way about a dozen persons were victimized. The messages were not all alike. Sometimes it was a house the general had rented or bought, with the amount correspondingly higher. Sometimes it was jewelry, and at other times something else. The swindlers were eventually caught, however, and severely sentenced.

A MODERN "ST. JOHN."

Some Cincinnati and Indianapolis merchants were similarly swindled by a gang of whom a former operator of the Western Union named James P. St. John. who afterward assumed the name of White, was a prominent member. Early one morning St. John called at the Western Union branch office, Third Street, Cincinnati. Being early, no one was present but the janitor. St. John represented himself as an employe of the main office, said he wanted to trace a message, and asked to see the messenger's delivery book. The book was afterward missing, and the matter reported to Manager Armstrong. It was not found until a week later, when the cashier of the Lafayette Bank called at the main office to ascertain whether a dispatch received from Saratoga, N. Y., signed "Springer," and ordering payment to Duhun & Co. of $450 for jewelry, was genuine. The dispatch was written on a regular No. 1 blank. Mr. Armstrong pronounced it a forgery. A short time afterward, when a young man presented the bill to the paying teller of the bank, he was arrested and imprisoned, and proved to be the same person who stole the messenger's book. Another forged dispatch, similar to the above, purporting to be sent by H. Hirsch, who was East, was delivered at the store of H. Hirsch & Co. same day, requesting them to pay Duhun & Co. $300 for goods previously purchased by him. In this case the swindlers were more successful. A confederate of St. John's shortly afterward presented a bill for $300 on one of Duhun & Co.'s billheads, and a check for the amount on one of the Cincinnati banks was given him. Instead of presenting the check at the bank he went direct to the establishment of Duhun & Co., where he represented himself as an employe of Hirsch & Co., gave a plausible excuse for the check being drawn in favor of Duhun & Co., selected seventy-five dollars worth of jewelry, and received $225 change, the check being pronounced genuine at the bank. Upon learning of the arrest of his confederate he left the city.

At Indianapolis a jeweler was similarly victimized out of $285 by St. John. The Cincinnati parties consenting, a requisition was procured for him, and he was conveyed to Indianapolis and tried, convicted, and sentenced to eight years' imprisonment. His confederate, who victimized Duhun & Co., and whose name has not transpired, was finally traced to Chicago, where he was arrested and returned to Cincinnati for trial. He had in his possession when arrested a number of telegraph blanks stolen at New York, Baltimore, and other points.

BIG SWINDLE IN TOLEDO.

Another example of swindling reaches us from Toledo, where a business firm, who are largely engaged in the grain trade, received what purported to be a dispatch from a correspondent named Wilson, at Jackson, Michigan, stating that there was a good opening at Dexter for purchasing wheat, and requesting the Toledo firm to send him $1,000 by American Express, and to notify him by telegraph when the money was sent.

A package containing the amount required was accordingly placed in the express office at Toledo, addressed to Mr. Wilson, Dexter, and a telegram also sent to Wilson, notifying him of the fact. About the same time the express agent at Dexter received a telegram from Jackson, signed Wilson, directing him to deliver the package to a man who would call for it, describing in the telegram minutely a man who afterward called, asked for, and received the $1,000 package. For a week or two the Toledo firm quietly awaited advices from Wilson in reference to his wheat purchases, and in the meantime the parties who had sent forged telegrams and obtained the money, felt so jubilant at their success that they told a confidential chum at Jackson how they had operated.

"SPIRITUALISTIC" SWINDLING.

A class of persons who live on the amiable credulity of the public, find the electric fluid a useful auxiliary. We mean the "spiritualists," so called, whose success in making money from the rich and ostensibly the cultured is no less remarkable than that of the advertising "clairvoyants," and the rest of the swindling sisterhood, who show a poor girl the portrait of her "future husband," for a "consideration" proportioned to the slender means of the ignorant victim. The fraud of "spiritualism" has not as yet been fully exposed, but enough has been discovered to make it plain that "mediums" are largely indebted for the manifestations they develop to the electric fluid.

MR. FAULKNER'S REVELATIONS.

Mr. Faulkner is a philosophical instrument maker, doing business in London. He writes that for many years he has had a large sale for spirit-rapping magnets and batteries, expressly made for concealment under the floor, in cupboards, under tables, and even for the interior of the centre support of large round tables and boxes; that he has supplied to the same parties quantities of prepared wire, to be placed under the carpets and oil-cloth, or under the wainscot and gilt beading around ceilings and rooms—in fact, for every conceivable place; that all these were obviously used for spirit rapping, and the connection to each rapper and battery was to be made by means of a small button, like those used for telegraphic bell ringing purposes, or by means of a brass-headed or other nail under the carpet, of particular patterns known to the spiritualists. He describes these rappers as calculated to mislead the most wary, and adds that there are spirit-rapping magnets and batteries constructed expressly for the pocket, which will rap at any part of the room. He has also made drums and bells which will beat and ring at command; but these two latter are not so frequently used as the magnets are, because they are too easily detected.

MAGNETS FOR "SPIRIT RAPPING."

A correspondent of the *English Mechanic's Magazine* has written an account of his methods of preparing apparatus for "spirit-rapping" meeting. We reprint it in his own words:

"In making my magnets for electric or 'spirit-rapping' drums I proceeded as follows: I took five bars of $\frac{1}{4}$ inch iron (one of them being very soft), 10 inches long, and filed them up. Around four of them I wound five layers of 32 silk-covered wire. Remember, the layers were complete, and all leading the current in the same direction. Around the fifth I put one layer. Of course the bars were bent into horseshoe shape. The magnets were bound together so as to bring the fifth or last as near as possible in the center, and its ends to project 1-64th inch beyond the others. A piece of zinc as thin as writing paper was next soldered on *one* pole of the centre magnet. Now for the keeper. It was made of a piece of soft iron 1-16th inch thick and about 3 inches square; one side of it had a half of a split lead bullet soldered to the centre. This gave the keeper weight, and prevented it from recoiling when it fell. I had three, and sometimes four, guide bars on my keepers; but I believe that, for all ordinary purposes, two are sufficient. These bars are made very smooth, and fitted into holes made in the brass framework supporting the magnets. The whole was now placed inside the drum. A word about this drum. In the first place, it should be a very common looking one; secondly, it should be—in fact *must* be—pretty large, say at least 2 feet in diameter—the larger the better. In fastening the 'electric drummer' inside, do so in such a way that it will not affect the sound. If your magnets are of good iron—that is, soft and without flaws—and well made, you will be able to work the keeper from a depth of half an inch, which, when it falls on the bottom of a large drum, will make a pretty loud thud. Now get two of those brass rings with the brass screws attached, used for boxes, &c., and fasten them through the woodwork in the top of the drum, and solder the collected ends of the magnet wire to them. Next close the drum up, and it is ready. Now, suppose you wish to amuse a number of people in your own rooms, you must find a way from the battery to the center of the room ceiling for the wires, so that they will be screened from observation. Let the wires terminate in two hooks to catch the drum-rings. By the bye, it looks less suspicious to hang the drum on three hooks, which you can easily do. You can use a battery of six pint Daniell's cells, and have a contact breaker in another room, to be attended to by a friend; or, if you can manage it, run the wires under the carpet, and work the contact with the heel of your boot, having a spring for raising the top wire when the pressure is off. Use one beat for 'no,' two for 'doubtful,' and three for 'yes.'

SIR CHARLES WHEATSTONE'S EXPERIMENTS.

This eminent gentleman exhibited some curious electrical experiments for the amusement of his friends, in which the developments were remarkably like those greedily devoured by the believers in spiritualism who patronize the magazines which support that delusion. We read that in a dark room, by a stamp of his foot, Sir Charles produced a brilliant crown of electric light in mid-air, while musical instruments seemed to be played by invisible hands; whereas the sounds really came from an adjoining room, in which the player sat, and, by an ingenious contrivance, were made to appear to be produced by the instruments before the spectators. A contest between science and the "spirits" in their own chosen feats would be almost as memorable as the celebrated competition between Moses and the magicians.

SHARP WORK BY OPERATORS.

The accounts we shall give under this head may not be thought, perhaps, to cast the same discredit or guilt upon the parties involved as in the foregoing; but the reader with the least moral sensibility cannot object to our use of the phrase "sharp work," although he might prefer the substitution of the adjective "smart" for the one employed. The first two tell the manner in which two poor operators became capitalists by the exercise of their abundant wit, to speak as gently as may be.

A youth of nineteen, who was a telegraph operator in Virginia City, on a salary of a hundred dollars a month, and who, when he could not make out German names in the list of San Francisco steamer arrivals, used to ingeniously select and supply substitutes for them out of an old Berlin city directory, made himself rich by watching the mining telegrams that passed through his hands, and buying and selling stocks accordingly, through a friend in San Francisco. Once, when a private dispatch was sent from Virginia, announcing a rich strike in a prominent mine, and advising that the matter be kept secret till a large amount of the stock could be secured, he bought forty "feet" of the stock at twenty dollars a foot, and afterward sold half of it at eight hundred dollars a foot, and the rest at double that figure. Within three months he was worth $150,000 and had resigned his telegraphic position.

Another operator, who had been discharged by his company for divulging the secrets of the office, agreed with a moneyed man in San Francisco to furnish him the result of a great Virginia mining lawsuit within an hour after its private reception by the parties to it in San Francisco. For this he was to have a large percentage of the profits on purchases and sales made on it by his fellow conspirator. So he went, disguised as a teamster, to a little wayside telegraph office in the mountains, got acquainted with the operator, and sat in the office day after day, smoking his pipe, complaining that his team was fagged out and unable to travel—and meantime listening to the dispatches as they passed over the wire from Virginia. Finally, the private dispatch announcing the result of the lawsuit sped along the wires, and as soon as he heard it he telegraphed his friend in San Francisco:

"Am tired waiting. Shall sell the team and go home."

This was the signal agreed upon. The word "waiting" left out would have signified that the suit had gone the other way. The mock teamster's friend picked up a large amount of the mining stock at low figures before the news became public, and a fortune was the result.

TAMPERING WITH CIPHER MESSAGES AND THE RESULT.

A San Francisco, California, newspaper gives the following interesting account of what came of tampering with cipher dispatches, in which it is shown that the operator and his friends did not in this instance fare quite so well as did the others above alluded to.

"The business office of the Chollar Mining Company is in San Francisco, and its works in Virginia City, Nevada. Correspondence between the superintendent at the latter place and the business office is kept up by both letter and telegraph, and, to prevent any inquisitive person from obtaining the contents of the telegrams in advance of their receipt by the officers of the company, a cipher was used. It had become apparent that certain brokers of San Francisco were regularly in receipt of reliable information concerning the condition of the mine, even before such information was obtained at the company's office. Just as soon as the superintendent in Virginia would send a cipher telegram stating that ore had been struck in any level or drift, these brokers would be on the street buying stock. Whenever he telegraphed bad news, they would appear as sellers at cash, or to deliver.

"That the trick was somewhere in the telegraph offices was evident, and to confirm this a plan was arranged, to which the superintendent, the office in San Francisco and the telegraph company were parties. The superintendent presented a cipher telegram, which, when interpreted, read after this style: 'Have struck the ledge; very rich; buy 3,000 shares if you can.'

"Although no one knew that this telegram was to be sent, and so far from the ledge having been struck the workmen had not been at work in the drift, yet, before this cipher was received at the San Francisco office, another telegram, addressed to the suspected brokers, had been sent and received, which contained precisely the same information and advice. On the strength of this these brokers rushed frantically out of their offices and commenced buying up Chollar stock at any price. In the Board they pursued the same plan, and finally loaded themselves with the stock, which rose in value as they bought, and sank when they ceased buying, their loss being estimated at between $15,000 and $20,000.

"A telegraph operator in the Virginia City office was immediately charged with having translated the cipher telegram, and upon the presentation of the evidence acknowledged his offence, and confessed the names of the brokers by whom he had been subsidized."

THE BITERS BIT.

The following shows how the best laid plans do not always bring the results that we desire:

During General McClellan's campaign in the Peninsula the gold and grain speculators of a certain city in a Northwestern State, organized an independent board or club, and had a wire run in from the Western Union Telegraph office.

The manager of the Western Union office soon became satisfied that there was a leak somewhere; for certain persons who did not belong to the club received the daily news sent to this branch office as soon as the parties to whom the dispatches were addressed, and speculated thereon. Investigation disclosed the fact that a meek looking young man, an operator in a country office, had been imported for the occasion; and, sauntering about the room with other outsiders, absorbed the contents of the dispatches, and instantly hied forth and communicated them to his employers.

Accordingly, having arranged a bogus dispatch, defeating McClellan with terrible slaughter, and sending gold up three or four per cent., the manager notified the *bona fide* subscribers not to act upon it, and sent it from the main office early in the morning.

The gentleman from the country swallowed it, and his friends bought gold *ad libitum* of the *bona fide* members, who chuckled at the trap they had caught the chaps in.

Great was the glee of the members of the board. They had caught the miscreants at last—and *wouldn't* they squeeze them!

When the regular dispatches were received, however, it was found that McClellan *had* been whipped! and gold *had* gone up, even higher than the bogus dispatch stated. *Tableau!*

The country operator retired, with his friends, on his share of the earnings, and the *bona fide* board was many thousands of dollars poorer.

A BANK SWINDLED BY BOGUS MESSAGES.

A gentleman who recently returned from a business trip to Texas, relates how a bank was swindled out of

$10,000 by three telegraph operators. It is highly improbable that there is much truth in the story, yet there is a bare possibility that such a scheme might be successfully carried out, and its publication may have the effect of putting banks and telegraph managers on their guard.

This gentleman says that one day a well-dressed man of business appearance presented at one of the banks in Dallas, Texas, a check for $10,000 on a well-known New York banking house, and desired it cashed.

He brought with him numerous letters of recommendation from persons with whom the bank had business transactions, and, so far as surface indication went, everything was right. But $10,000 was a considerable sum to pay out, even on the very best documents of recommendation, and the bank officers hesitated, wavered, and finally declined to cash the check. But the stranger was importunate. "Gentlemen," said he, " I came to Texas to invest this money in cotton. It is very necessary that this check should be cashed or I will be greatly inconvenienced. Suppose you telegraph to New York to this banking house? Ask them about me; I will pay all expenses."

Nothing could be more plausible than this; nothing sound more honest. So a dispatch was sent asking about the stranger and the check, and in a short time came the answer to the effect that it was all right, and the Dallas Bank would confer a favor on the New York firm by accommodating their cotton speculative friend and cashing the check. Still the bank officers were not satisfied, and another dispatch was sent. Again the answer was of a similar tenor, only probably a little more emphasis was added to it. This was satisfactory, and the check was duly cashed.

When at night the Dallas office, as usual, came to compare the number of messages sent during the day with the number received from it by the several offices with which it was in communication, it was found that neither of the dispatches sent by the bank had been received at the office to which they should have gone, and consequently no answers could have been sent. It was evident that the bank had been swindled, but how? There was the mystery. The dispatches had been regularly received; they had come from somewhere, but where from could not be known. The cotton speculator had disappeared with the funds, and the bank officials were at their wits' ends.

In a day or two the mystery was solved. Two operators, who had been employed in the Dallas office, and had resigned on the day before the well-dressed stranger made his appearance, had gone a few miles out of Dallas, taken possession of an old shanty by the roadside, attached an instrument to the wires, and taken off the dispatches intended for New York. They had then sent pre-arranged answers. The three were confederates, and the operators knew about the time the bogus speculator would enter the bank, and when to attach the instruments. It was an adroit scheme and successfully carried out. The bank got no clew to the swindlers, but learned a valuable lesson, paying a high price for tuition.

TELEGRAPHIC TRAP FOR BURGLARS.

This chapter will be fitly concluded with an account of a device to catch safe-burglars, the invention of a Mr. Barb, of London, who has patented it. The depredator no sooner begins to force open the door, drill the lock, or move the safe, than by so doing he sends a telegraphic message to the nearest police office, exhibiting the number of the safe he is attacking; and this number, registered in the police-books, has opposite to it the address of the house in which the robbery is being effected. The invention is a very simple thing. An instrument termed the "communicator" is fitted inside the safe; it consists of a small bolt, which is forced back upon a coil-spring when the door is closed, and which, in opening or moving the door, is instantly set in motion. In connection with this bolt wires are led through the bottom or the back of the safe and concealed in the wall, or inclosed within gas or water pipes, and, communicating with the street telegraph wires, are connected with the "alarm" and indicator at the police-station. The effect of tampering with the door or other part of the safe is to sound the alarm-bell at the police-station, and to exhibit on the face of the instrument the number of the safe. Arrangements are, of course, made to obviate sending of alarms on ordinary and legitimate occasions of using the safe, by simply putting the apparatus out of gear at the pleasure of the owner. The simple operation of turning a small key is all that is required to render the wires available, after which the owner may leave his premises, perfectly confident that electricity will keep a tireless watch over the property left in its custody.

THE TELEGRAPH AN UNIVERSAL INSTITUTION.

We need scarcely say to the intelligent reader of these pages that the use of the telegraph may now be said to be universal throughout the world.

As an illustration of this universality, we may cite the transmission of a telegraphic message sent by Courtney, the Auburn, New York State, oarsman, to Trickett, a brother in his profession, then resident in Australia. This message was sent from Auburn to New York City, thence to Heart's Content, Newfoundland, the cable end, thence to Valentia, on the coast of Ireland, thence to London, then through Germany, Russia, Siberia, thence to Wladiwodstock, a point on the coast of Manchuria; thence through Japan Sea to Nagasaki, on one of the Japan Islands, through the Yellow Sea, to Shanghai, China, thence down the coast of China through China Sea to Taigon, Siam, to Singapore, Malay, thence to Batavia, on the coast of Java, thence to St. Darwin, on the northern coast to Australia, and lastly to Sydney.

Many interesting things are told of the introduction of lightning as a servant in countries which do not rank high in the possession of that civilization which may be characterized as of the nineteenth century.

SUPERSTITION IN SPAIN.

Not long ago a London newspaper published an account from a town called Lorca, in Spain, described as containing twenty thousand people, and a thriving commercial centre. The people in the neighborhood of this place firmly believe in the existence of certain wizards—mysterious beings, with pale faces and long white beards, who, hid during the day, hunt at night for children, whom they devour. The fat of these children they are said to keep sacredly for two purposes—first, as a sovereign cure for small pox; and, secondly, to grease the wires of the electric telegraph, which is in itself a satanic invention, and would not work at all were it not for the lubricating oil obtained from the bodies of innocent little children.

MOROCCO.

After this who will be surprised to learn that upon the introduction of the electric telegraph into Morocco it was vehemently opposed by many who looked at the progress of the work with religious horror? The emperor threatened with death any person who should injure the apparatus, but the inhabitants of the little village of Mahovany, nevertheless, cut down the wires. The irate emperor straightway had the place surrounded by his troops, and the heads of ten prominent citizens were forthwith cut off and fixed on the telegraph poles, as an awful warning.

CHINA.

The first telegraph (telephone) line in China was six miles in length, and erected about two years ago by Li Hung Chang, viceroy of China, from his official residence to the Tietsen arsenal. There was no attempt at interference by the native populace, as in the case of telegraphs projected by foreigners; but it is stated that the people were afraid of the apparatus, thinking that little devils run along the wire and carry the messages. In consequence of this superstition they had previously torn down a few lines put up by foreigners. We may add that such outrages do not now attend the erection of telegraphs in the Celestial Empire.

INDIA.

When the electric telegraph was established by the English in India, its introduction was accompanied with curious and difficult problems. In the first place, it was discovered that the air of India is in a state of constant electrical perturbation of the strongest kind, so that the instruments there mounted went into a high fever, and refused to work. Along the north and south lines a current of electricity was constantly passing, which threw the needles out of gear, and baffled the signalers. Moreover, the tremendous thunder-storms ran up and down the wires, and melted the conductors; the monsoon winds tore the teak-posts out of the sodden ground; the elephants and buffaloes trampled the fallen lines into kinks and tangles; the Delta aborigines carried off the timber supports for fuel, and the wire or iron rods upon them to make bracelets and supply the Hindoo smitheries; and the cotton and ice boats, kedging up and down the river, dragged the subaqueous wires to the surface. In addition to these graver difficulties were many of an amusing character. Wild pigs and tigers scratched their skins against the posts in the jungle, and porcupines and bandicoots burrowed them out of the ground. Kites, fishing eagles and hooded crows came in hundreds and perched upon the line to see what on earth it could mean, and sometimes were found dead by dozens, the victims of their curiosity. Monkeys climbed the posts and ran along the lines, chattering and dropping an interfering tail from one wire to another, which tended to confound conversations with Calcutta.

EARLIEST TELEGRAPHS IN THE EAST.

One of the earliest telegraph lines in Eastern countries was a private line erected in 1859, from Teheran to Sultanieh, where the shah of Persia temporarily resided. This line, one hundred and sixty-nine miles long, after being used one summer, was abolished. Of the construction of the line from Shahrud to Meshed, the Persian inspector-general of telegraphs reported: "The workmen suffered very much from want of water and from heat. During the two months of June and July, 1876, the heat in the plains, with quite a cool wind blowing, rose to 140° Fahrenheit, while the heat in the shade once rose to 112° Fahrenheit. Great anxiety was felt on account of the Turcomans, who were expected to attack us every day, but not a single Turcoman was seen." The first through telegraph to the far East was erected by the Turkish government in 1863, and extended from Constantinople through Asia Minor, by way of Mosul, to Bagdad. In 1864 the government of British India built a line on iron standards, from Bagdad to Fao, at the head of the Persian Gulf. This line was subsequently handed over to the Turks, and was deemed so unsafe, passing as it did through a region where the Porte had really little or no authority, that after the submarine cable from Fao to Kurachee had been laid, a telegraph line was put up by British officers, but at the cost of the Persian government, from Bushire *via* Theran to Bagdad.

The proclamation by which the king of Burmah announced his intention to construct a system of telegraphy for the use of his subjects, is a curious example of Oriental official literature. It intimated that the "present Founder of the City of Mandalay or Rutapon, Builder of the Royal Palace, Ruler of Sea and Land, Lord of the Celestial Elephant and Master of many White Elephants, Owner of the Sekyah or Indra's Weapon, Lord of the Power of Life and Death and Great Chief of Righteousness, being exceedingly anxious for the welfare of his people, in the year 1213 will introduce the telegraph—a science—the elements of which may be compared to thunder and lightning for rapidity and brilliancy, and such as his royal ancestors, in successive generations, had never attempted to subdue."

JAPAN.

The Japanese take kindly to occidental innovations, the leading men in the empire interesting themselves in a most commendable manner to advance its civilization. Our own country has taken a very creditable share in the introduction of improvements among that singular nation of the far East, which, resembling its neighbors the Chinese in many respects, is directly the opposite in gladly welcoming every innovation which is, or promises to be, an improvement. The American government was the first to initiate the "Japs" in the operations of the field telegraph, by presenting one to the mikado, in the imperial presence. By the mikado's desire the apparatus was erected in the grounds of the palace, one terminus being his majesty's private study, and the other the pleasure pavilion which stands in the center of the Maple Gardens, where were assembled three princes of the blood, the prime minister, and a host of members of the privy council, to receive and answer the imperial messages. The working of the wires was entrusted to two Japanese, and when all was ready a message arrived at the pavilion announcing the presence of the mikado at the terminus in the study. To this announcement a most respectful message was returned, thanking his majesty for his gracious presence.

Shortly afterward the message came: "The emperor is highly pleased with the wonderful Western invention," and then immediately followed: "Who are in the pavilion, and what are you doing?" To this an answer was returned, giving the names of those present, and saying that they were waiting with profound veneration his majesty's gracious orders. To their intense embarrassment the next thing heard was: "Telegraph to us something amusing." As may be imagined, this message caused the greatest consternation among the courtiers. How were they in a moment to conjure up anything that should be amusing, and, at the same time, respectful? At length one privy councillor suggested: "This day will be memorable in the annals of the empire as that on which his majesty for the first time witnessed the working of a telegraph." But this was instantly rejected as being not in the least amusing. At last a youthful courtier proposed: "We all mean to get merry on the wine which we expect your majesty to give us." This was at once received with delight, and transmitted to the palace; and to it a reply was immediately returned that they should not expect in vain, and the proceedings terminated with a message from the emperor expressing himself satisfied with the experiments, and thanking the officers who had worked the telegraph. At the emperor's desire the apparatus was left standing in the grounds, in order that he might learn to work it himself.

AFRICA.

Nothing has been more remarkable in the history of the last few years than the progress of discovery in the continent of Africa, which promises to shortly open it up fully to the operations of trade, aided by the steamboat, the locomotive, and the electric telegraph. A correspoudent of the London *Times*, writing recently from Berba, in tropical Africa, says:

"It was singular to meet with the telegraph in the heart of the desert between Aryab and Berba; not the telegraph put up and in working order, as we see it in Europe, but all the appurtenances of that instrument of civilization carried on the backs of hundreds of camels, which, laden with coils of wire and hollow iron posts, trod their toilsome path through the burning sand. Every now and then we met one of these poor beasts which, overweighted and broken down by the weight of his load, had fallen on the ground and been abandoned a victim to the vultures. All this telegraphic gear was marked "Siemens Brothers, London," and was *en route* to Khartoum, from which town it will be forwarded on to span the desert between Kordofan and Darfour. A good many lives will probably be sacrificed before the line can be considered open, as the Arabs, who eagerly steal every piece of iron they can meet with for their spear points, have to be very severely punished before they leave off cutting down the poles. However, this difficulty once got over, the telegraph will be as easily worked as the one between Khartoum and Cairo, which, when it was first laid down, was continually being interrupted."

Thus the march of improvement steadily progresses, and the dark places of the earth are being provided with the agencies which enlarge and refine life.

THE WEATHER REPORTS.

The recent death of Brigadier-General Albert J. Meyer, chief signal officer of the army, gives painful interest to a subject with which his name was long identified, one, moreover, of the greatest importance to the interests especially of our commercial marine and of agriculture.

STORM SIGNAL SYSTEM.

We are indebted to the pen of the deceased gentleman for the best account of this system, written with singular clearness, exactness and completeness. The following passage occurs in one of General Meyer's annual reports, addressed to the secretary of war. He says:

"Synchronous observations are taken and forwarded three times daily, at about 8 A.M., 6 P.M. and 12 midnight, by careful observers, under military control, and supplied with the best instruments, namely, barometer, thermometer, hygrometer, anemometer and rain gauge. The observations are forwarded by telegraph, in the shape of a numeral cipher, the intelligence conveyed in sixty words being sent in a twenty-word report.

"The telegraphic transmission of the regular reports has presented a problem difficult of solution. The list of stations of observation and report exhibits a large number of stations, so located that if reports are to be both received from and sent to them two or three times a day without an organization of working especially designed for the purpose, the delays would be great, and the repetitions, each of which involves a chance of error, numerous.

"The extensive lines of the Western Union Telegraph Company and the co-operating companies, the International Ocean Cable Company and the Northwestern Telegraph Company, have been divided into circuits. These circuits reach in their course every station of observation and report. Each circuit thus provides for a certain group of stations. This being arranged, the working forms of circuits set forth minutely the telegraphic labor needed for the movement of the messages of each group; for the exchange of message reports between different groups—between different places in different groups; and, finally, for the assembling of all the dispatches in Washington."

WHAT THE SIGNAL SERVICE DOES FOR COMMERCE AND AGRICULTURE.

What specific purposes, it may be asked, are answered by the department over which General Meyer so ably presided? "The Signal Service, United States Army, Division of Telegrams and Reports for the Benefit of Commerce and Agriculture," is expected to perform the following duties, detailed by the same accurate pen as the foregoing quotation: "To give protection to commerce by warnings on all of the Atlantic and Gulf coasts of the United States, and on those of the lakes; to watch the river changes along their courses in the great river valleys; to note at seasons the temperatures affecting canal commerce; to carry telegraphic lines, by which meteorological reports may be had, over regions considered impracticable for such constructions; to maintain a system of connected stations on the seacoast; to take charge of the recognized system of voluntary meteorological observations on this continent, in addition to the regular system of the service; to secure the co-operation of foreign observers in foreign countries; to endeavor to aid directly all the farming population in the harvesting of their crops; and, finally, to put it in the power of every citizen to know each day, with reasonable accuracy, the approaching weather changes."

We need not add that this gigantic intention has been and is carried out with a degree of efficiency which is surprising, and which is continually increasing.

In order to insure its accomplishment, there is a thorough course of instruction given to those who are to be observers, both in military signaling and telegraphy, meteorology and the Signal Service duties at stations of observation and report. This is done at the school of instruction and practice at Fort Whipple, Virginia. Upon their being found efficient, the observers are placed at stations where, in such of these as forward telegraphic reports, "they are required to take, put in cipher, and furnish, to be telegraphed tri-daily on each day, at different fixed times, the results of observations made at those times, and embracing in each case the readings of the barometer, the thermometer, the wind velocity and direction, the rain-gauge, the relative humidity, the character, quantity and movement of upper and lower clouds, and the condition of the weather." In addition to the reports supplied to the daily papers, what are called Farmers' Bulletins are furnished daily to such post offices as can be reached from convenient centers.

THE NEW YORK STATION.

The New York station of the Signal Service is situated on the top story of the building occupied by the Equitable Life Assurance Company. It commands a superb view of the city and bay, and affords a place for the display of the cautionary signals where they are visible from all parts of the harbor. The lantern, displaying a red light, is one hundred and ninety-five feet above sea-level; while the flag—red, with a black centre—floats from a staff at an elevation of two hundred and thirty-five feet.

Here is the wind vane, which needs no description, and also the anemometer, used to determine the velocity of the wind. By electric connections with ingenious but not complex machinery, this is a self-registering instrument. The train of wheelwork makes and breaks an electric circuit, which registers itself on the paper revolving by clockwork on the recording cylinder.

There is in the New York office a self-registering

barometer. It is a rare and splendid instrument. One of the cylinders, which are revolved by clockwork, gives the register of the barometric changes for a day, and the other for a period of fifteen days. As in the anemometer, the connections between the instrument itself and the recording cylinders are made by electricity.

Both to save time and expense, as well as to insure accuracy, the telegraphic reports of the service are made in cipher. These ciphers are easily and quickly read by means of a book arranged for the purpose. Here, for example, is the cipher report of the observation taken at New York on a certain day: "York, Monday, Dead, Fire, Grind, Himself, Ill, Ovation, View;" which, translated, reads:

York : New York (Station).
Monday : 30.07 (Barometer corrected).
Dead : 29.90 (corrected barometer for temperature and instrumental error).
Fire : 70° (thermometer).
Grind : 75 per cent. (humidity).
Himself : west, fair (wind and weather).
Ill : 6 miles (velocity of wind).
Ovation: ½ cirrus clouds, calm (upper clouds).
View: 67° (minimum temperature during night).

The Signal Service is an exacting one. From the chief officer down to the privates, to the men in comfortable quarters in the cities and to the men who winter on Mount Washington or Pike's Peak, the thanks of the whole community are due for their tireless service.

EARLY OPPOSITION.

The system we have described, and which has proved so successful that the proportion of failures is now less than ten per cent., was not adopted without opposition. No less a man than Mr. A. Watson, of Washington, described as the originator of the idea of storm signals, wrote to a New York journal that the plan of telegrams and reports, then just adopted, had been abandoned by the Smithsonian Institution. Speaking of the storm signal system, he wrote:

"In furtherance of this plan of telegrams and reports the department has enlisted fifty sergeants as meteorologists, at $900 per annum, making $45,000, which, added to the $15,000 appropriated by Congress, makes $60,000 at least, to be expended this year, which same reports were obtained through the telegraph company by the Smithsonian Institution at no cost whatever. But fifty sergeants are as yet employed as meteorologists, and stationed at different parts of the country to telegraph the weather, which number may, perhaps, be increased to hundreds if not thousands, costing a million of dollars or more per annum. The Western Union Telegraph Company has three thousand five hundred operators throughout the country, which, at $900 per annum, would amount to $3,150,000. And every one of these, by my plan or by any other, will have to be employed to telegraph storms and floods, or else employ sergeants in equal number. But why employ a sergeant to inform a telegraph operator of the state of the weather, or that a storm is passing in a certain direction, when that agent can know it as well as the other, and has control of the wires to telegraph on all sides and to any distance to assure himself of the certain extent, direction and intensity of the storm or flood? These gentlemen are as intelligent as any that can be found, and for a trifle additional compensation would do the work. It is plain that the station agents located at the principal towns are all the meteorologists that are needed, and are the only persons that can do the work complete and at a trifling cost.

"What is needed is a sound signal, by cannon, to give instant and general warning, for many miles in all directions, of coming storms and floods. I predict that these weather reports will prove a total failure and a costly one at that. As the great War Department and the portentous Signal Office have been seven months in devising and putting into operation these weather reports, it reminds me of the old but apt saying, 'The mountain was in labor and brought forth a mouse,' and in this instance the mouse is very little and old at that."

The well-informed reader remembers that equally severe remarks were made at the expense of the locomotive upon its first introduction, not to speak of other gigantic improvements which were made at the cost of influential opposition, but, like our storm signal system, soon justified their existence by their beneficent results.

ORIGIN OF WEATHER REPORTS IN THE UNITED STATES.

Mr. Watson's letter naturally leads to an inquiry as to what had been done in the way of furnishing weather reports to the people of this country previously to the formation of an army department provided for that purpose. This subject was treated with success by Professor Cleveland Abbé, in the August number (1871) of *The American Journal of Science*, according to which the first published suggestion of the feasibility of weather reports appears to be that of Professor William C. Redfield, in *The American Journal of Science* for September, 1846, where he states that "in the Atlantic ports the approach of a gale may be made known by means of the Atlantic telegraph, which probably will soon extend from Maine to the Mississippi." The next mention of the subject is found in the Smithsonian report for 1847, in an article by Professor Elias Loomis, who wrote: "When the magnetic telegraph is extended from New York to New Orleans and St. Louis, it may be made subservient to the protection of our commerce, even in the present imperfect state of our knowledge of storms." But however frequently the idea may have been suggested of utilizing our knowledge by the employment of the electric telegraph, according to Professor Abbé, it is to the late Professor Henry, of the Smithsonian Institution, that the credit is due of having first actually realized this suggestion, as acknowledged by the Vienna Academy of Sciences.

The practical utilization of the results of scientific study is well known to have been greatly furthered by the labors of this institution, and from the very beginning Professor Henry successfully advocated the feasibility of telegraphic storm warnings. It will be interesting to trace the gradual realization of the earlier suggestions of Redfield and Loomis in the following extracts from the annual Smithsonian reports of the years indicated :

1847. "The extended lines of telegraph will furnish a ready means of warning the more northern and eastern observers to be on the watch for the first appearance of an advancing storm.

1848. "As a part of the system of meteorology, it is proposed to employ, as far as our funds will permit, the magnetic telegraph in the investigation of atmospherical phenomena. * * * The advantage to agriculture and commerce, to be derived from a knowledge of the approach of a storm by means of the telegraph, has been frequently referred to of late in the public journals—and this we think is a subject deserving the attention of the government.

1849. "Successful applications have been made to the presidents of a number of telegraph lines to allow us, at a certain period of the day, the use of the wires for the transmission of meteorological intelligence. * * * As soon as they (certain instruments, etc.) are completed, the transmission of observations will commence." (It was contemplated to constitute the telegraph operators the observers.)

1850. "This map (an outline wall map) is intended to be used for presenting the successive phases of the sky over the whole country at different points of time, as far as reported."

1851. "Since the date of the last report the system, particularly intended to investigate the nature of American storms immediately under the care of the Institution, has been continued and improved." The system of weather reports thus inaugurated continued in regular operation until 1861, when the disturbed condition of the country rendered impossible its further continuance. Meanwhile, however, the study of these daily morning reports had led to such a knowledge of the progress of our storms, that in the report for 1857 Professor Henry writes:

1857. "We are indebted to the National Telegraph line for a series of observations from New Orleans to New York, and as far westward as Cincinnati, which have been published in *The Evening Star* of this city. We hope in the course of another year to make such an arrangement with the telegraph lines as to be able to give warnings on the eastern coast of the approach of storms, since the investigations which have been made at the institution fully indicate the fact that, as a general rule, the storms of our latitude pursue a definite course."

Before peace had been proclaimed, after the civil war, Professor Henry sought to revive the systematic daily weather reports, and in August, 1864, at the meeting of the North American Telegraph Association, a paper was presented by Professor Baird, on behalf of the Smithsonian Institution, requesting the privilege of the use of the telegraph lines, and more especially in order to enable Professor Henry "to resume and extend the Weather Bulletin, and to give warning of important atmospheric changes on our seaboard." In response to this communication it was resolved "That this association recommend * * * to pass free of charge * * * brief meteorological reports * * * for the use and benefit of the institution." Upon the communication of this generous response, preparations were at once made for the undertaking, and its inauguration was fixed for the year 1865. In January of that year, however, occurred the disastrous fire which seriously embarrassed the labors of the Smithsonian Institution for several years. It became necessary, therefore, to indefinitely postpone the work, which indeed had through its whole history been carried on with most limited financial means, and was quite dependent upon the liberal co-operation of the different telegraph companies.

It will thus be seen that without material aid from the government, but through the enlightened policy of the telegraph companies, and with the assistance of the munificent bequest of James Smithson, "for the increase and diffusion of knowledge," the Smithsonian Institution organized a comprehensive system of weather reports, which, although since, as we have shown, superseded by one more complete and efficient, ought still to be held in grateful remembrance and be accorded a proper acknowledgment.

THE RAILWAY TELEGRAPHIC SYSTEM.

One of the great daily papers of New York city published, some time ago, an article assuming that moving trains by telegraph was an American institution, and gave a detailed account of its first application in support of this assumption.

AN UNFOUNDED ASSUMPTION.

The article stated that "the first practical application of telegraphic signals in moving trains was made on the Erie line in 1850." It added that "previous to that time locomotive engineers and conductors were distrustful, and there are several instances on record of their positive refusal to obey telegraphic orders, especially when their trains were directed to proceed beyond stations, to meet and pass trains going in opposite directions, except in cases where such orders were plainly expressed in printed orders upon their regular time tables. In 1850, however, when the Erie road had but a single track between Piermont and Elmira, it was plainly demonstrated to the superintendent (the late Charles Minot) that the telegraph would be a very important assistance to the road, and it became plainly evident that the telegraphic service must eventually be adopted upon all main trunk lines.

"When the first telegraphic message was sent over the Erie wires a train filled with western bound passengers was lying at Turner's Station, awaiting the arrival of an eastern bound train, which, by the time table, should meet and pass it at that point; but, owing to an accident two hundred miles west, it could not possibly arrive until five or six hours later. Mr. Minot was a passenger upon the train lying at Turner's. He immediately decided to test the accuracy of the telegraph, and make a beginning of the plan of ordering trains to proceed to points further in advance, and not further delay the stationary train, when the track was known to be clear as far as Port Jervis, a distance of one hundred and fifty miles further west. Orders were accordingly sent over the wire to the station agent at Port Jervis to hold all easterly bound trains until the arrival of the western train. This order was given in order to make all safe, and prevent a collision in case the former should arrive at Port Jervis before the latter. An answer was immediately given by the station agent, announcing that he fully understood the order, and would do as directed. All appeared safe, and the engineer was ordered to start west; but, to the astonishment of Mr. Minot, he positively refused to move the train from Turner's upon any such arrangement. Mr. Minot immediately mounted the locomotive, pulled out the throttle valve, and ran the train himself, assisted by the fireman, and reached Port Jervis according to programme.

"The ice was broken, and since that time the telegraph has been acknowledged as a positive necessity on all long railroad lines in this country. As many as twenty trains have since moved in opposite directions at one time upon a single division of the Erie road with perfect safety. The form of giving the necessary directions, however, has been somewhat changed; and now the conductors and engineers of each train who receive telegraphic directions are telegraphed the name of the particular point at which they are to meet, and answers are required from them, to ascertain whether they understand orders, before any movement is made."

TRAIN DISPATCHING AN ENGLISH INVENTION.

However gratifying this account may be to our national pride, and useful as embodying in its last sentences information of current value, so far as it pretends to give an account of the introduction of train dispatching, it is not to be trusted. Charles H. Haskins, now general superintendent of the Northwestern Telegraph Company at Milwaukee, and prominently connected with telephone matters in that section, when he was conductor upon the Michigan Southern Railroad, in the winter of 1849-50, telegraphed to hold a boat at Monroe for his train, which had been detained by an accident. This is probably the first instance of a train order on our side of the Atlantic. There is no doubt that the English were the first to adopt telegraphic signaling on railways. An English pamphlet entitled "Telegraphic Railways; or, the Single Way Recommended by Safety, Economy and Efficiency, under the Safeguard and Control of the Electric Telegraph, &c. By Wm. Fothergill

Cooke, Esq."; published in London in 1842, has a large chart, illustrating fully the manner in which trains were to be moved on a single track by means of telegraphic signals or orders, given by the station masters from station to station. The instruments then in use on the Blackwall Railway are illustrated by diagrams, and the use of these instruments fully explained.

GRAND CENTRAL DEPOT SIGNAL SYSTEM.

At the Grand Central Depot, Forty-second Street, New York City, are the termini of three great railroads, and here the telegraphic signal system is carried to such a height of perfection as to merit particular description. With the exception of the interval between 1:10 and 3:40 in the morning, and of fifty minutes at noon, no period of fifteen minutes elapses in which some train does not depart or arrive *via* the Harlem, the Hudson River, or the New York, New Haven and Hartford road. One hundred and eighteen regular, and from ten to fifteen extra trains daily pass in one direction or the other over the tracks on the underground road between **Fifty-Third Street** and **Harlem Bridge**, a distance of nearly four and a half miles. Barely two minutes sometimes intervene between the departure of one train and the incoming of another, and three trains often start at intervals of five minutes apart.

It is obvious that, in order tó prevent confusion and accident, the movements of each and every one of these trains, while traveling between the points named, must be governed with absolute certainty. Add to this that crowd after crowd of passengers must be admitted from the reception room to the outgoing cars at exactly the proper time, and the checking of their baggage must be stopped in time to insure its dispatch by the proper trains; and the reader will have formed some faint idea of the perfect system which must exist for the management of the machinery of the great depot and its approaches.

Located far up on the north wall of the depot, the view from its broad window extending over the intricate network of rails into which the various tracks diverge, is a small cabin. On the wall hang signal indicators and bells, time-tables, and a huge clock. On the table before the single occupant are a telegraph instrument, a record book, and three rows of ivory buttons, twenty in all. This is the dispatcher's office, and here, by pressing the buttons or manipulating the telegraph key, he controls the movement of every train going or coming, the buttons, though simple electric bells, governing everything near and about the depot, the key transmitting instructions to far-off points. By way of illustration, we suppose that one train is to start at 4:30, and that another will arrive at 4:31 o'clock. It is now just 4:10, the passengers are congregated in the waiting-room, the cars are in place, and the engine, with steam up, is standing outside not yet attached. The dispatcher touches a button, the sound of a bell is heard, the heavy doors of the waiting room fly open, and the passengers crowd upon the cars. Fifteen minutes elapse; the operator presses another button, a gong strikes in the baggage room, and the checking is stopped. Belated individuals who wish to depart by that train must go *minus* their baggage. Now the operator watches the clock closely; three minutes pass, and then a sharp peal rings out from a bell close beside him. The minute hand points to 4:28, and the incoming train has reached Sixty-fourth Street and is signaling its own approach. The sound continues for half a minute, then stops; the train is at Fifty-fifth Street, and the finger of the dispatcher at once presses another button. If we were on the arriving locomotive we would see a green disk before us, or at night the flash of a green light, meaning that everything is ready for the flying switch just outside the depot, by which the engine is to clear itself from the train, the cars entering the depot by their own momentum. Now it is 4:29; down goes another button; a bell on a post beside the locomotive waiting outside rings for the engineer to back in and couple on. Hardly ten seconds elapse before a sharp "ting" calls the operator's attention to the fact that the pointer arm of the indicator on the wall has swung over from "clear" to "block." The arriving train is on the Fifty-third Street crossing. The clock says 4:30; again a button is pressed; the doors of the waiting-room are slammed shut, there is a few seconds' delay for the tardy ones on the platforms to board the cars, and then the train moves slowly out of the depot.

The indicator pointer still shows "block," and if the outgoing train continues its course a disastrous meeting on the crossing may result. The dispatcher remains passive, however, for he knows that the signal between that train and the crossing is normally at "danger," and that the engineer will certainly come to a stop and wait until the red disk is turned. The delay is but for a second, for the indicator bell almost instantly sounds again, the arm swings over to "clear," and the proper button is immediately touched. A distant cloud of steam can be seen for a moment, and the outgoing train is off again. Pressing another button the operator restores the danger signal. The arriving train now rushes in, its passengers disembark, and at the sound of the bell from the dispatcher, a locomotive kept for the purpose couples on and drags the empty cars out of the depot.

We have accounted for twenty-one minutes, during which one train has left and one arrived; the reader may imagine the celerity and certainty of the work when we add that, within fifteen minutes spent in the dispatcher's cabin, three trains on three different roads were started and three received, all at different times and without the slightest confusion.

MOVING TRAINS BY TELEGRAPHIC ORDERS.

As the above account does not, of course, cover all the information properly coming under the head of this chapter, a statement of the general system, given in detail, cannot fail to be exceedingly interesting. The following are the instructions issued to the employes of a prominent railroad:

"Superintendents and train dispatchers are the only persons who are authorized to move trains by special orders. Before an order is given by telegraph for two or more trains to meet at a given station, the red signal to stop the trains must first be displayed at such meeting point; and until this is done no order must be sent to either train. When a meeting or passing point is to be made by two or more trains, the order must be definite and conclusive, and sent first to the conductor having the right to the road. If it is desired to give a train the right to run against a passenger train, the order is first sent to the conductor of the latter, and no order must be given the opposing train until the receipt of a satisfactory reply from the conductor of the passenger train. And in the same way, before giving a passenger train the right to the road, over a train possessing this right, the order should first be sent to the train holding the right to the road, and when a satisfactory reply has been received from the conductor of the train, then the order may be transmitted to the other train. All special orders for the movements of trains, whether sent by telegraph or otherwise, must be communicated in writing. When a train is abandoned, the order of the superintendent directing its abandonment must be sent by telegraph to all agents, conductors and engineers on the division.

"No train must leave a station to run upon the time

of an abandoned train, which by the regulations would have the right of the road, unless the conductor and engineer of such trains have in their possession a copy of the order of abandonment properly signed and certified to by the operator. If a train should be held at night at any telegraph station where there is no night operator, the conductor must call the day operator into his office for the purpose of receiving the orders necessary before going ahead. At stations where telegraphic orders are awaiting an expected train, operators will display a red flag by day, or a red light by night. When orders are duplicated to following trains, the understanding of each conductor and engineer must be separately written, and must be responded to by the party giving the order."

THE TRAIN DISPATCHER AND OPERATOR.

A moment's reflection makes it manifest that the position of train dispatcher is second in importance to no position on the road. He is frequently found to be also an expert operator, but it is not absolutely necessary that he should be a telegrapher. It is his duty to keep the localities of every train running on the division constantly in his mind, and issue orders to them at every station where they stop. Where the roads have only single tracks, the labors and responsibility of a train dispatcher are very great, sometimes as many as twenty trains, traveling in opposite directions, being on the division at one time. All these have to meet and pass each other somewhere along the division. The dispatcher must know just where to hold the train, where to send that one from and how far to run it, and know within a second just when to expect a train at a station. With his time card before him, containing the names of all stations and numbers of all trains, the dispatcher sits close to the operator, surrounded by clicking instruments, checks off train and station as arrivals are rapidly telegraphed, and quickly issues his orders to the operator, to be sent to expectant trainmen all along the division. The dispatcher of trains on a single track is the player of a gigantic game of chess, the men in which are to be so moved that they may never be brought in check. For any accident by collision on a road, the dispatcher is held responsible, unless it is shown that his orders were disobeyed.

His companion in incessant vigilance should not be overlooked. One of them writes with a sprightliness which is surprising: "Imagine yourself stranded at an out-of-the-way station, right in the woods as likely as not, and nothing more exciting than the monotonous train report, with its 'Os, Os, No. 3 X O. T. at 9:15 As,' and the everlasting string of figures and ciphers in the car-report day in and day out, with now and then a variation in the shape of a wreck, which keeps all hands up all night sending crossings and repeating orders back to dispatchers with the mystic 'I understand,' etc., and see how you relish the prospect. Then, by way of thanks from an appreciative public, comes the newspaper report in the case of every other accident or so, censuring the telegraph and saddling the whole blame on some defenceless operator, who has been so badgered and worried and overworked that he couldn't swear to the difference between an x and a g if he were to be hung for it."

A THRILLING INCIDENT.

The following instance illustrates the responsibility attached to the position of train dispatcher:

The chief dispatcher upon a prominent western road had ordered Miss D., operator at M. station, to "hold the through freight bound east for further orders." The sharp spiked staff, bearing its warning signal flag, was set in its usual conspicuous place, but just as the expected train rounded the curve a treacherous puff of wind blew the flag to the ground, unnoticed by the operator, whose horror can be more easily imagined than described as, a moment after, the "through freight," with a shriek and a roar, swept past the station and its unseen and therefore unheeded signal, while a few miles ahead, on the same single track, an extra was speeding along in the opposite direction, under orders that had been given by the dispatcher after the signal had been set at M. Miss D. hurriedly notified the dispatcher of the accident. There was one more telegraph station for the through freight to pass before meeting the extra, and upon the chance of the faithfulness of this one operator was hung the fate of the rapidly approaching trains. While the dispatcher's operator " called " the station rapidly and incessantly, the dispatcher himself walked the floor in an agony of suspense, and Miss D. could see nothing, hear nothing but the rapid clicking of the instrument burning itself into her very brain and nerves. The moments passed like hours. For some minutes there was no response, but at last came the welcome "aye, aye." One moment more of breathless suspense while the question is put: "Has engine No. — passed?" "Not yet." "Thank God!" exclaims the dispatcher, and gives the necessary orders to avert the impending calamity.

THE ANGUISH AND SUSPENSE OF AN OPERATOR WHO "FORGOT."

Fortunately it is seldom that an operator, however harassed he may be with other cares and responsibilities, neglects to deliver the orders he has received for a train. That he should occasionally slip in this particular would be no more than could be expected from mortal beings who have more to look after than any one man should. At many stations the operator, in addition to his regular telegraphic duties, is called upon to sell tickets, check baggage, attend to express and freight matters, and the like, which the public in their intercourse with him too often seem to entirely overlook. Whenever a case of neglect in connection with a train order does occur, an example is made of the man by instantly dismissing him. An operator who was thus discharged writes as follows of his feelings after he had found that the train had gone by which he should have signaled to stop:

"I forgot, and in doing so have forfeited the respect

of my employers as well as my position. I am not one who is in the habit of forgetting, and to forget when scores of lives depend on my memory and carefulness makes me shudder when I think what might have come of my forgetting this time.

"It was the first time in my life that I had made a serious mistake. I received orders for the east-bound passenger train yesterday to look out for a freight train ahead of them to the next station east of me, and I forgot to put my signal flag out to stop the passenger train. Operators know what my mistake means. It means, sometimes, death to the unfortunate passengers, but, thank God, in this case no harm came of it, for the freight train had a good start and got safely in on the side-track before the passenger train came along. Minutes were years to me while waiting for the signal from the other station that would decide whether all were safe or not. My heart was in my mouth every time the line would open, and with fear I listened to every stroke of the sounder. At last the welcome "rep." G. came, and I knew everything was safe. I could no longer bear up under the enormous strain that my nerves had been subjected to for fifty minutes. I raised myself from the instrument table, staggered to the door, and but for the fresh air would have fainted. My wife saw me and was frightened to see my pallid face. The cold perspiration stood out on my forehead, my hands trembled as with palsy, and my breath came in gasps as I tried to regain possession of myself.

"Those who have passed through this ordeal can imagine my feelings; the awful dread which comes over one when the discovery is first made that a train has passed which should have been stopped. Hideous, laughing demons dance before one's imagination, in seeming mockery of the anguish that is dragging one almost to madness. In imagination you see the mangled and lifeless forms that but a moment ago passed you, full of happiness and radiant in anticipation of meeting dear ones at their destination.

"The remorse, the utter helplessness that overtakes one in such a moment is indescribable. It is a time the memory of which will haunt me through life. Amid the joys and pleasures that may await me in the future, there will ever be a spot as black as midnight darkness. My hand trembles when I take hold of the key that may deal death and destruction to the many lives entrusted to my care."

A NOBLE RAILROAD OPERATOR.

An account has reached us from Parker City, Pennsylvania, of an operator whose presence of mind and firm persistency saved probably many lives. She was employed at Sligo Junction, and one evening was awaiting the arrival of an eastern-bound train, when a long freight, numbering fifty-five heavily laden cars, bound westward, slacked up at the station, which the conductor entered. After registering he shouted "all right" to his engineer, and was about to get on his train when the operator, remembering something caught from the subtle click of her instrument, rushed out and bade him stop his train, which was slowly moving away. He replied that they had the right of way. "No," said the operator, "I heard an order passing over the wire, telling you to remain here until the arrival of the Sligo passenger train, which you will certainly meet, because it is somewhere between this and the next station westward. This order should have been given you at Troy" (a station just passed by the freight). The conductor was persuaded to enter the office by the operator, who, going to the key, asked the superintendent at Brookville if such an order had been sent to Troy. "Yes," was the reply. The careless operator at Troy had failed to deliver this important order, but the carefully trained ear of the female in charge at Sligo Junction had caught it, and the result was the hasty switching of that long freight from the main track to the side. Just as this was done a shrill whistle announced the coming of the Sligo train, the headlight of which threw a glare along the track.

Thankful were the passengers, conductor, engineer, and brakemen that their lives were in the hands of one who fully realized the fact. "For," said the engineer of the freight, "I could not have stopped my heavy train in less than a mile. Our escape would have been impossible." The conductor and engineer of the passenger train feelingly caught the operator's hand and thanked her for preventing an accident which would have certainly cost them their lives. The superintendent at Brookville telegraphed his thanks, and in a few days sent a letter of commendation, inclosing a bank check, with the thanks of the officers of the road, as a mark of their appreciation of her care and attention to business.

NEW INVENTIONS IN RAILROAD SIGNALING.

An invention has been secured by patent in this country, by a Swede, whose apparatus is an automatic railway signal which enables the station officers to know the precise position of any train at any time; it gives sound signals to the engineer and at the station before the train enters, thus enabling switches to be cleared and arranged in time to prevent accidents. If two trains approach each other, whether running in the same or opposite directions, the engineers of both trains receive signals in time to prevent collisions, and the station people are at the same time automatically informed of the position of both trains. Any train may be stopped at certain points on the road where "contacts" are arranged upon telegraphic communication with the stations at both ends of the route, and two trains may in the same manner telegraph to each other. A complete record is automatically kept at each station of the speed of each train, and of the exact time it enters or leaves the station. Stop-signals may be sent at any time from the stations to any train while moving. The apparatus may be arranged to send stop or danger signals to trains approaching swinging bridges which are not properly locked and fastened.

The Union Electric Signal Company of Boston, Massachusetts, lately exhibited the practical working of a new method of automatic railway signals, known as Robinson's Contact Circuit Rail system. The experiments were made at a street-crossing of the Boston and Providence Railroad, near Boston, in the presence of a number of gentlemen interested in American railway matters. The system tested on this occasion differs essentially from other systems of signaling in use, in that the rails instead of wires are employed for conducting an electric current. They are divided into sections, according to curves and other contingencies. At one end of each section is placed a small battery, one pole of which is connected to either rail, and at the other end of the section is an electro-magnet, the coils of which are connected to the two rails, thereby establishing a constant metallic circuit through the rails and magnet. At either end of the section is the standard bearing the signal, which is connected with the circuit. When a train enters upon the section, the leading wheels and axle of the engine instantly short-circuit the current, the magnet is demagnetized, and on the latter leaving its armature the signal is mechanically thrown to danger, where it remains as long as the wheels are on the section, and when they pass off the signal goes back to safety, and the section of the line is open to receive another train. By this means the rear of a train on a road equipped with these signals will always be safely guarded. Actual experience has demonstrated that the rails are vastly superior as conductors to any surrounding media, and that the electricity will adhere to them in preference to passing off to earth, despite rain or snow. After the operation of the system had been witnessed for upward of an hour, as the various up and down trains passed the crossing, it was pronounced unanimously to be an unqualified success, and calculated to be of great service in perfecting the safety appliances which all the American railways will ere long be compelled to use.

Right here is the place to say that a Swiss inventor envelopes the driving axle of locomotives in coils of insulated copper wire, and by the passage of an electric current converts the wheels into powerful magnets with increased adhesion to the rails.

A new system of telegraphic signals has been introduced by way of experiment at the Boston end of the Lowell Railroad. A box in the train house of the passenger depot on Causeway street is connected by wires with the office of the ticket master, who, when a train starts, by the pressure of a finger upon a little instrument, displays at an aperture in front of the box a red flag if in the day, or if at night time the red sides of a lantern to view. When the engine reaches the rail directly in front of the station at East Cambridge the electric current is opened, and the red flag disappears or the lantern shows a white light. A bell is also rung at the same moment in the ticket office. If this system works well, and is adopted along the line generally, it may easily be so perfected that knowledge of the position of a train may be known at any station which it has just left, and thereby insure comparative safety to passengers.

Many accidents have occurred from trains breaking apart, the engineer not being notified of the fact. There has long been needed some connection throughout the train more effective than the old-fashioned bell-rope, which, though perhaps sufficient for passenger trains, is not applicable to freight. Major V. B. Bell has brought out an invention, especially adapted to freight trains, which promises to secure the desired end. It is simply a train telegraph. In one corner of the caboose is a battery, differing from common telegraph batteries in being constructed of leather and copper, and being closely boxed—connecting with an alarm in a small box on the side of the caboose and with another on the engine; wires run beneath all the cars, and the connection is established between the cars by flexible copper wires, covered, which can be detached, being held in their places by any single spring catch—the common spring clothes-pin being used at present. When the train breaks, these cords are unfastened, the connection is broken, the alarm is sounded in the caboose, and the engine and the train is stopped. This is the principal object of the invention; but by means of it the conductor can, by simply moving the key of the alarm box, signal the engineer to back, go ahead, etc. A thorough test of it was recently made by practical railway operatives and managers, and the results are pronounced satisfactory.

Though the machinery was necessarily imperfect—being all new and untried—the inventor was able to answer all objections and explain how all proposed difficulties may be easily surmounted. The apparatus would cost about seventy-five dollars.

Nervous people will appreciate the announcement that locomotive whistling promises to be an abomination of the past. At Poughkeepsie, in this State, it is to be superseded by a bell worked by electricity, which will be set up in the depot. When the train arrives within a mile of the station, the bell will ring until it gets to the depot. The danger signal is thus given, and the waste of steam is avoided, to say nothing of the racket.

It may not be entirely out of place to close this sub-section with the statement that an interesting application of electricity, in connection with a tank for supplying locomotives with water, is now in operation at Buda Station, on the Chicago, Burlington and Quincy Railroad. The steam pump which supplies the tank is on the bank of a small stream half a mile distant, and entirely out of sight. A float is arranged so that if the water is drawn off to a level more than two or three inches below the top of the tank a circuit is closed, connecting by wires with the pump house. This sets an alarm bell ringing within hearing of the engineer, who then starts his pump, and runs it till the tank is full, of which due notice is given by the cessation of the alarm. This arrangement was devised by Fred. H. Tubbs, then superintendent of telegraphs on the C. B. and Q. R. R., but now superintendent of the American Union Telegraph Company at Chicago, and has worked for a long time in the most satisfactory manner.

FUN ABOUT SIGNALING.

With the American propensity to relieve overtaxed energies with harmless nonsense, everybody is familiar, as all commend it. The subject of this chapter has proved most fertile in nonsensical suggestions and funny yarns, but we resist the temptation to enlarge it excepting by a little of such material.

ROUSING THE SLEEPING CAR PORTER.

From the West we hear of a gentleman lately returned to Milwaukee from a trip, who tells of a new use that has been found for electricity that even beats the telephone or the phonograph. It is a device by which the colored sleeping car porter can be awakened at every station. It is well known that the normal condition of the colored person is to be asleep. The colored person goes to sleep on the slightest provocation. In the ordinary affairs of life this eccentricity can be overlooked and provided for, but the business of sleeping car porters has baffled scientists to devise a method of keeping them awake. A porter can be kept awake by constantly whistling, but this practice has a tendency to awaken passengers who do not desire to be awakened. The inventor has adapted electricity to this branch of railroading in such a manner that the colored person's usefulness is increased, at very little expense. It is desirable that the porter should be awake at each station where the sleeper stops, in order to snatch the small baggage of those who get aboard, and throw it under the seats.

A wire runs from the engine under the cars, and is connected with an electric disk in the cushion of the hind seat of the sleeper, where the colored man is apt to congregate, and at the same moment that the engineer rings the bell on approaching a station, he touches the thingumbob attached to the wire. Suppose the porter to be seated in his accustomed place, peacefully dreaming the happy hours away. His head is thrown back, his eyelids are in repose, his mouth is open like an approach to a tunnel. He is sitting on the electric disk. The hand of the engineer playfully touches the cornucopia, the lightning flashes back to the sleeper, a charge of electricity goes meandering up the spinal column of the African, he is raised toward the roof of the car, and when he comes down he is wide awake and ready for business.

ANOTHER ANTI-SLEEPING INVENTION.

It is said that at a certain station on the Philadelphia and Erie Railroad, the company has a new night telegraph operator who, if inclined to slumber, is too ingeniously wide awake to be caught napping at his post. Recently he was seized with drowsiness, which he could not shake off. As it was his duty to report all passing trains, he dared not yield, and yet could not resist. That mother of invention, necessity, at length suggested an alarm signal, which he proceeded to put in operation by suspending a scuttle full of coal by means of a cord which was passed through the keyhole of his office door, and fastened across the track at the requisite elevation. Mr. Operator then resigned himself to rosy dreams, which were finally interrupted by a passing train, the engine of which snapped the cord, causing the coal-scuttle to come down with a rattle-te-bang that would have aroused even a sleeping New York policeman. Another young operator, some thirty miles up the road, let a train slip by him the same night, and applied to the inventor of the coal-scuttle alarm to know when the train passed his station. No answer was vouchsafed, the inventor remarking: "Why don't the blockhead get the right to use my patent?"

ELECTRICITY AND LIFE.

Very little is known of electricity, perhaps it may truly be said nothing beyond what has been observed of its effects. What it is in itself, its relations or possible oneness with heat and light, are unknown. Professor Faraday, on one occasion, in speaking on the nature of electricity before the British Association for the Advancement of Science, thus expressed his views: "There was a time when I thought I knew something about the matter; but the longer I live and the more carefully I study the subject, the more convinced I am of my total ignorance of the nature of electricity." Enough of its operations have been seen and noted, however, to suggest that its possible usefulness is beyond present calculation and even conception. Among these is the influence it possesses in the stimulation and support of both animal and vegetable life, including the highest development of the first named, so far as we know, in the human being. Every reader is, of course, acquainted with the fact that the electric battery is extensively employed as a remedial agent, and that experiments are constantly in progress with the view to determine, if possible, the exact value of electricity in therapeutics, and in the case of children, animals and plants. A brief statement of selected information on the general subject will be both interesting and valuable.

THE ELECTRIC GIRL OF LA PERRIERE.

The extraordinary phenomena we are about to relate occurred in the commune of La Perriere, situated in the department of Orne, France, in January, 1846. They seem to be properly authenticated, and are not incredible in themselves. The astonishing electrical force exerted by the electric eel, found in some rivers of South America, is familiar to everybody, and shows the force which can be employed by the animal organism when charged, so to speak, with the electric fluid.

Angelique Cottin, a peasant girl fourteen years of age, robust and in good health, but very imperfectly educated and of limited intelligence, lived with her aunt, the widow Loisnard, in a cottage with an earthen floor, close to the chateau of Monti-Mer, inhabited by its proprietor, M. de Faremont.

The weather for eight days previous to the fifteenth of January, 1846, had been heavy and tempestuous, with constantly recurring storms of thunder and lightning, and the atmosphere was charged with electricity.

On the evening of that fifteenth of January, at eight o'clock, while Angelique, in company with three other young girls, was at work as usual in her aunt's cottage, weaving ladies' silk-net gloves, the frame, made of rough oak and weighing about twenty-five pounds, to which was attached the end of the warp, was upset and the candlestick on it thrown to the ground. The girls, blaming each other as having caused the accident, replaced the frame, relighted the candle and went to work again. A second time the frame was thrown down Thereupon the children ran away, afraid of a thing so strange, and, with the superstition common to their class, dreaming of witchcraft. The neighbors, attracted by their cries, refused to credit their story So returning, but with fear and trembling, two of them at first, afterward a third, resumed their occupation, without the recurrence of the alarming phenomenon. But as soon as the girl Cottin, imitating her companions, had touched her warp, the frame agitated again, moved about, was upset, and then thrown violently back. The girl was drawn irresistibly after it, but as soon as she touched it, it moved still further away.

Upon this the aunt, thinking, like the children, that there must be sorcery in the case, took her niece to the parsonage of La Perriere, demanding exorcism. The curate, an enlightened man, at first laughed at her story; but the girl had brought her glove with her, and fixing it to a kitchen chair, the chair, like the frame, was repulsed and upset, without being touched by Angelique. The curate then sat down on the chair; but both chair and he were thrown to the ground in like manner. Thus practically convinced of the reality of a phenomenon which he could not explain, the good man reassured the terrified aunt by telling her it was some bodily disease and, very sensibly, referred the matter to the physicians.

The next day the aunt related the above particulars to M. de Faremont; but for the time the effects had ceased. Three days later, at nine o'clock, that gentleman was summoned to the cottage, where he verified the fact that the frame was at intervals thrown back from Angelique with such force that, when exerting his utmost strength and holding it with both hands, he was unable to prevent its motion. He observed that the motion was partly rotary, from left to right. He particularly noticed that her feet did not touch the frame, and that when repulsed she seemed drawn irresistibly after it, stretching out her hands as if instinctively toward it. It was afterward remarked that when a piece of furniture or other object thus acted upon by Angelique was too heavy to be moved, she herself was thrown back, as if by the reaction of the force upon her person.

On the twenty-first of January the phenomena increased in violence and in variety. A chair on which the girl had attempted to sit down, though held by three strong men, was thrown off, in spite of their efforts, to several yards distance. Shovels, tongs, lighted firewood, brushes, books, were all set in motion when the girl approached them. A pair of scissors fastened to her girdle was detached and thrown into the air.

On the twenty-fourth of January, M. de Faremont took the child and her aunt in his carriage to the small neighboring town of Mamers. There, before two physicians and several ladies and gentlemen, articles of furniture moved about on her approach. And there, also, the following conclusive experiment was tried by M. de Faremont:

Into one end of a ponderous wooden block, weighing upward of a hundred and fifty pounds, he caused a small hook to be driven. To this he made Angelique fix her silk. As soon as she sat down and her frock touched the block, the latter was instantly raised three or four inches from the ground; and this was repeated as many as forty times in a minute. Then, after suffering the girl to rest, M. de Faremont seated himself on the block, and was elevated in the same way. Then three men placed themselves upon it, and were raised also, only not quite so high. "It is certain," says M. de Faremont, "that I and one of the most athletic porters of the Halle could not have lifted that block with the three persons seated on it."

Dr. Verger came to Mamers to see Angelique, whom, as well as her family, he had previously known. On the twenty-eighth of January, in the presence of the curate of Saint Martin and of the chaplain of the Bellesme hospital the following incidents occurred. As the child could not sew without pricking herself with the needle, nor use scissors without wounding her hands, they set her to shelling peas, placing a large basket before her. As soon as her dress touched the basket, and she reached her hand to begin work, the basket was violently repulsed, and the peas projected upward and scattered over the room. This was twice repeated, under the same circumstances. Dr. Lemonnier, of St. Maurice, testifies to the same phenomenon, as occurring in his presence and in that of the procurator royal of Mortagne; he noticed that the left hand produced the greater effect. He adds that he and another gentleman having endeavored, with all their strength, to hold a chair on which Angelique sat down, it was violently forced from them, and one of its legs broken.

On the thirtieth of January, M. de Faremont tried the effect of insulation. When, by means of dry glass, he insulated the child's feet and the chair on which she sat, the chair ceased to move, and she remained perfectly quiet. M. Olivier, government engineer, tried a similar experiment, with the same results. A week later, M. Hebert, repeating this experiment, discovered that insulation of the chair was unnecessary; it sufficed to insulate the girl. Dr. Beaumont, vicar of Pin-la Garenne, noticed a fact, insignificant in appearance yet quite as conclusive as were the more violent manifestations, as to the reality of the phenomena. Having moistened with saliva the scattered hairs on his own arm, so that they lay flattened, attached to the epidermis, when he approached his arm to the left arm of the girl, the hairs instantly erected themselves. M. Hebert repeated the same experiment several times, always with a similar result.

M. Olivier also tried the following: With a stick of sealing-wax which he had subjected to friction, he touched the girl's arm, and it gave her a considerable shock; but on touching her with another similar stick that had not been rubbed, she experienced no effect whatever. Yet when M. de Faremont, on the nineteenth of January, tried the same experiment with a stick of sealing-wax and a glass tube, well prepared by rubbing, he obtained no effect whatever. So also a pendulum of light pith, brought into close proximity to her person at various points, was neither attracted nor repulsed in the slightest degree.

Toward the beginning of February, Angelique was obliged for several days to eat standing; she could not sit down on a chair. This fact Dr. Verger repeatedly verified. Holding her by the arm to prevent accident, the moment she touched the chair it was projected from under her, and she would have fallen but for his support. At such times, to take rest, she had to sit herself on the floor, or on a stone provided for the purpose.

On one occasion, "she approached," says M. de Faremont, "one of those rough, heavy bedsteads used by the peasantry, weighing, with the coarse bedclothes, some three hundred pounds, and sought to lie down on it. The bed shook and oscillated in an incredible manner; no force that I know of is capable of communicating to it such a movement. Then she went to another bed, which was raised from the ground on wooden rollers, six inches in diameter; and it was immediately thrown off the rollers." All this M. de Faremont personally witnessed.

On the evening of the second of February, Dr. Verger received Angelique into his house. On that day and the next upward of one thousand persons called to see her. The constant experiments, which on that occasion were continued into the night, so fatigued the poor girl that the effects were sensibly diminished. Yet even then a small table brought near to her was thrown down so violently that it broke to pieces. It was of cherry-wood and varnished.

"In a general way," says Dr. Beaumont-Chardon, "I think the effects were more marked with me than with others, because I never evinced suspicion, and spared her all suffering; and I thought I could observe that, although her powers were not under the control of her will, yet they were greatest when her mind was at ease and she was in good spirits." It appeared, also, that on waxed or even tiled floors, but more especially on carpets, the effects were much less than on an earthen floor like that of the cottage where they originally showed themselves.

At first wooden furniture seemed exclusively affected; but at a later period metal also, as tongs and shovels, though in a less degree, appeared to be subjected to this extraordinary influence. When the child's powers were the most active, actual contact was not necessary. Articles of furniture and other small objects moved, if she accidentally approached them.

Up to the sixth of February she had been visited by more than two thousand persons, including distinguished physicians from the towns of Bellesme and Mortagne and from all the neighborhood, magistrates, lawyers, ecclesiastics and others. Some gave her money. Then, in an evil hour, listening to the mercenary suggestion, the parents conceived the idea that the poor girl might be made a source of pecuniary gain; and notwithstanding the advice and remonstrance of her true friends—M. de Faremont, Dr. Verger, M. Hebert and others—her father resolved to exhibit her in Paris, where the phenomena continued for a time and then ceased.

Dr. Tanchon says that a chair which he held firmly with both hands was forced back as soon as she attempted to sit down; a middle-sized dining-table was displaced and repulsed by the touch of her dress; a large sofa, on which Dr. Tanchon was sitting, was pushed violently to the wall as soon as the child sat down beside him. The doctor remarked that when a chair was thrown back from under her, her clothes seemed attracted by it, and adhered to it until it was repulsed beyond their reach; that the power was greater from the left hand than from the right, and that the former was warmer than the latter, and often trembled, agitated by unusual contractions; that the influence emanating from the girl was intermittent, not permanent, being usually most powerful from seven till nine o'clock in the evening, possibly influenced by the principal meal of the day, dinner, taken at six o'clock; that when the girl was cut off from contact with the earth, either by placing her feet on a non-conductor or merely by keeping them raised from the ground, the power ceased, and she could remain seated quietly; that, during the paroxysm, if her left hand touched any object, she threw it from her as if it burned her, complaining that it pricked her, especially on the wrist; that, happening one day to accidentally touch the nape of her neck, the girl ran from him crying out with pain; and that repeated observation assured him of the fact that there was, in the region of the cerebellum, and at the point where the superior muscles of the neck are inserted in the cranium, a point so acutely sensitive that the child would not suffer there the lightest touch; and, finally, that the girl's pulse, often irregular, usually varied from one hundred and five to one hundred and twenty beats a minute.

These curious phenomena, which were given in the *Atlantic Monthly* in the year 1866, created great interest.

A case very similar to that of Angelique Cottin occurred in the month of December previous, in the person of a young girl, not quite fourteen years old, apprenticed to a colorist in the Rue Descartes, Paris. The occurrences were quite as marked as those in the Cottin case. The professor, seated one day near the girl, was raised from the floor, along with the chair on which he sat. There were also occasional knockings. The phenomena commenced December 2d, 1845, and lasted twelve days.

A WESTERN ELECTRICAL LADY.

The case of an American lady, resident at Nevada City, is interesting to scientific men, and not less so to those of us who remember our boyish freak of producing electric phenomena by rubbing poor pussy's coat headwards. For many years, we are told, the subject of this paragraph was afflicted with acute neuralgic pains in various parts of the body, and, hoping to find relief, resorted to the use of an electrical battery. She used the apparatus for six months, but found no relief. At that time nothing was noted of unusual character as the result, and although several months elapsed, it was only when cold weather commenced that any extraordinary symptoms followed. One night after this the lady had occasion to enter a dark room and pick up a woolen coat that was lying there. As she did she was both surprised and frightened to observe a bright light surrounding the hand that held the garment. At the same time the electric currents passed along the arm, shocking her quite severely. When her husband was informed of the fact he discredited its reality, thinking there was more imagination than anything else in it. So the next evening, to convince the incredulous better half, she turned the gas out in the room where they were sitting, and letting her hair down began combing it. A remarkable display of light was the result. The sparks flew around in every direction, and there was a sharp, cracking sound as the teeth of the comb passed between the hairs. In laying her hands upon iron the lady did not observe the peculiarities referred to; but the instant she touched a woolen cloth the fire began to fly, and the shocks followed one another in rapid succession.

ELECTRICITY ON THE DINNER TABLE.

An experimentist, Dr. Gladstone, says that in daily life weak electrical currents are at work where their presence is often little suspected: for instance, supposing a person at dinner to have a silver fork in one hand and a finger upon the steel part of a knife held in the other, it follows that, when he plunges the knife and fork into a beefsteak, two dissimilar metals are thereby placed in a moist conducting substance, consequently a voltaic circuit is formed, and an electric current flows through the body of the individual between the knife and fork. To prove that this was really the case, he connected a reflecting galvanometer with the knife and fork by means of wires; he then proceeded to cut a beefsteak, and the current thus generated deflected the needle of the galvanometer, so that the spot of light which it reflected was seen traveling along the screen by all the observers.

FEELING PULSE BY TELEGRAPH.

While lecturing, some time ago, Dr. Upham, of Salem, Massachusetts, in order to explain to his audience the variations of the pulse in certain diseases, caused the lecture room to be placed in telegraphic communication with the city hospital of Boston, distant fifteen miles, and by means of special apparatus the various pulse beats were exhibited by a vibrating ray of magnesium light upon the wall. These experiments have since been repeated at Paris with success.

DEVELOPMENT OF YOUTH BY ELECTRICITY.

Dr. Poggioli recently read a paper at a meeting of the British Academy of Medicine, on the "Physical and Intellectual Development of Youth by Electricity." He remarked that De Candolle had quoted experiments to show that vegetation is much richer and quicker in its growth when electrified than otherwise. Seeds subjected to the action of this fluid would yield better produce than others, and in a shorter time. Starting from these data, Dr. Poggioli conceived the idea that a similar action might be proved to exist in the animal kingdom, and especially in the case of young subjects. He adduced five instances of children, varying between the ages of four and sixteen, and having all attained a remarkable development, both in a physical and an intellectual sense. Among these there was a child which might be considered a phenomenon of deformity and stupidity, and that under the influence of electricity grew three centimetres in a single month, and has since been always first, instead of last, in his class. From this Dr. Poggioli concludes that the electric fluid exercises a direct influence over the physical and intellectual development of young subjects.

ELECTRICITY IN SURGERY AND DENTISTRY.

It is stated that when General Kilpatrick returned from Chili, a few years ago, he had a remarkable operation performed by a physician in New York, who removed a large fleshy formation from the general's neck by filling it full of needles and then attaching a galvanic battery to it. Ten minutes after the current of electricity was let on, the bunch had entirely disappeared.

Again, we learn that a Philadelphia dentist has invented a little machine for driving the filling into teeth, which works by electro-magnetism. The hammer, or "plunger," working within a small cylinder, may be made to deliver its blows at the rate of several hundred strokes a minute—so rapidly, indeed, as almost to produce the impression of a continuous pressure. A battery large enough to work the apparatus costs for running it about twenty cents a day.

ELECTRICITY IN MEDICINE.

At a recent sitting of the French Academy of Sciences, M. Scoutetten sent in a paper on certain further researches of his for the purpose of proving that the electrical state of mineral waters is the chief cause of their activity. He contends that these waters, on issuing from the earth, are in a state of peculiar activity owing to certain chemical reactions which produce dynamic electrical phenomena; a fact which by no means impairs the activity of their chemical elements on the human body.

A French physician says that a shock of electricity given to a patient dying from the effects of chloroform immediately counteracts its influence and restores the patient to life.

Already employed to restore vigor and nimbleness to the gouty limbs of decrepit *bons vivants*, the recent discoveries of Dr. Bernier, a French physician, show electricity to be an efficient remedy for the evil effects of excessive drinking on the human nose. The doctor maintains that, by the application of an electric current to noses even of the most Bacchanalian hue, the flesh may be made "to come again as the flesh of a little child;" and he supports his assertion by a case performed on a female patient of his own—a woman of high rank.

In connection with these instances of the value of electricity as a healer, we may fitly introduce the anecdote told of an elderly woman who entered a railroad carriage at one of the Ohio stations, and disturbed the passengers a good deal with complaints about a "most dreadful rheumatiz" that she was troubled with. A gentleman present, who had himself been a severe sufferer with the same complaint, said to her: "Did you ever try electricity, madam? I tried it, and in the course of a short time it cured me." "Electricity!" exclaimed the old lady, "yes, I've tried it to my satisfaction. I was struck by lightning about a year ago, but it didn't do me a single mossel o' good."

ELECTRICITY AN "ANTI-FAT" REMEDY.

We do not vouch for the accuracy of what we are about to relate, which records a remarkable operation performed by a Whitehall, New York State, physician. A gentleman who had been suffering from a superabundance of adipose tissue consulted a medical practitioner, asking for relief from his burden. The latter took him to the telegraph office at that place. The fat man was requested to remove his coat and vest, after which the physician surrounded him with wires, attaching the ends to a powerful galvanic battery. At a signal from the doctor, the manager let on the current. The patient writhed and twisted when he felt the current passing around him, but he stood it like a martyr. Presently he began to shrink; he grew smaller and smaller and smaller; his clothing hung in bags about his fast diminishing form. The doctor felt much pleased at the result of his experiment, while the formerly fat man's joy was very great, although he seemed to be suffering the worst pain. All of a sudden there was heard a loud clicking at the instrument, as if Pandemonium's great hall had been let loose. The operator sprang quickly to answer the call. He ascertained it was from the New York office, and quickly asked: "What's up?" An answer came back as if some demon was at the other end of the wire: "Cut off your wires quick—you are filling the New York office with soap grease!"

OUTGROWTHS OF THE TELEGRAPH.

This is a subject remarkably fertile, because never in the history of the world have there been the same incentives to, and, we may add, the same success in many-sided invention as now. How multifarious, for example, are the applications of steam power! But those of the electric telegraph are, perhaps, even more numerous and certainly more interesting by reason of their diversity and marvellousness. We shall cite some of these, taking care to give due prominence to the most important, but not pretending, within the limits yet remaining to us, to include the mention of every realized or projected employment of electricity in the industrial arts. The subject of this chapter has, moreover, been more or less anticipated under appropriate headings in the foregoing pages.

THE ELECTRIC LIGHT.

One of the best, if not the best, descriptions of the electric light is that given by Mr. Edison in the October number (1880) of the *North American Review*.

EDISON'S DESCRIPTION OF IT.

Mr. Edison begins the article with a few words to those who have expressed their impatience at the delays in the perfecting of the electric light. The delays which have occurred to defer its general introduction are chargeable, he says, not to any defects since discovered in the original theory of the system in its practical workings, but to the enormous mass of details which have to be mastered before the system can go into operation on a large scale, and on a commercial basis as a rival of the existing system of lighting by gas. Important improvements have been brought about by these delays, in the direction of economy and simplification at almost every point in the system, as well as in the details of manufacturing the apparatus.

The lamp, the inventor tells us, has been completely transformed. To quote his own words:

"The perfected lamp consists of an oval bulb of glass about five inches in height; pointed at one end, and with a short stem three quarters of an inch in diameter at the other. Two wires of platinum enter the bulb through this stem, supporting the loop or U-shaped thread of carbon, which is about two inches in height. The stem is hermetically sealed after the introduction of the carbon loop. At its pointed end the bulb terminates in an open tube through which the air in the bulb is exhausted by means of a mercury pump till not over one-millionth part remains. The tube is then closed. The outer extremities of the two platinum wires are connected with the wires of an electric circuit, and at the base of a lamp is a screw by which the circuit is made or broken at pleasure. When the circuit is made the resistance offered to the passage of the electric current by the carbon causes the loop to acquire a high temperature and to become incandescent; but as this takes place in a vacuum, the carbon is not consumed. The 'life' of a carbon loop through which a current is passed continuously varies from seven hundred and fifty to nine hundred hours.

With an intermitted current the loop has an equal duration of life, and as the average time an artificial light is used is five hours per day, it follows that one lamp will last about six months. Each lamp costs about fifty cents, and when one fails, another may be easily substituted for it."

In conclusion, Mr. Edison promises the speedy introduction of his perfected lamp. Meanwhile it is satisfactory to reflect that in many places the public are already in the enjoyment of a light which is a source of comfort, safety and beauty. The world does not wait even for Mr. Edison.

THE ELECTRIC LIGHT AT NIAGARA FALLS.

Recent experiments at the Falls of Niagara, which at this writing take place once or twice a week, give emphasis to the possibilities of beauty afforded by the electric light. Esthetic people will be delighted to learn that in these experiments the fantastic displays of color surpass the richest pigments of the painter. From the terrace at Prospect Park on a dull night the Falls appear, under the rays of a red electric light, like an immense and swift-sliding avalanche of purple lava. In a moment, by changing the stained glass in front of the electric lamp, the Falls are made to gleam like silver, and when alternate colors are employed, the appearance of a splendid moving rainbow is presented. The foam in the abyss at the foot of the Falls when lit by the electric glow shines out like the phosphoresence of the ocean during a tempestuous night. One of the most rare and striking scenes it is possible to witness is the sudden illumination of Niagara by a flash of night lightning, and with the electric light it will be possible to produce the effect artificially.

EXPERIMENTS IN SAN FRANCISCO.

As everybody knows, considerable portions of probably every great city in the leading countries of the world are now lighted by electricity, either as the result of private enterprise or municipal provision. In this connection it is announced that San Francisco claims to be ahead of European cities in the quality of the lamp used in lighting some of her streets. The light produced is said to be so brilliant that it cannot be looked upon with the naked eye without dazzling and injuring that delicate and most sensitive organ, it being even less painful to gaze upon the sun. One wonderful feature of this light is that any and every color is easily seen; the colored threads in various fabrics, the bright green of the grass, and the colors of flowers were brought out as distinctly as in daylight. By an ingenious device a light can be made self-supplying for the longest night. It is self feeding, and can be burned as long as desired. Twelve jars and a coil are required for each light, save where two are in the immediate neighborhood of each other, as on either side of a hill, when one set of jars and one coil will answer for both. The plan and ingredients are kept a profound secret, but the inventors claim that they can light the city for one hundred thousand dollars a year, which is only one-third the present cost of gaslight.

ELECTRIC LIGHTING ON AN EXTENSIVE SCALE.

It is reported that a Boston electrical engineer is about to try the experiment of lighting the large manufacturing center of Holyoke, Massachusetts, with the electric light in a manner that will strike the present generation as novel, but which has been essayed before. It is proposed to erect a tower seventy five feet high overlooking the town. This is to be surmounted by an immense lantern of such illuminating capacity as to put all previous lamps in the category of trifles. At present only one tower will be erected, but if the principle should prove a success, seven or eight will ultimately be built, with a view to render the city as light as day, and completely to supersede gas and kerosene. The idea of the inventor of this daring scheme is to charge the upper strata of the atmosphere with luminous vibrations in the same manner as is done by the sun, and thus to produce the same effect that is obtained during the day from the reflected, refracted and diffused light of that orb. In this manner it is believed that electric light can be made to permeate spaces which are inaccessible to direct rays by the same law by which daylight diffuses itself—that is by virtue of an expansive property which is constantly illustrated on the large scale of solar illumination, but has no place in our text-books on optics. The light given by the solar orb a few minutes after sunset, when only the upper strata of the atmosphere are directly affected by the solar beam, furnishes, perhaps, the best example of the diffusion and expansion that the engineer proposes to imitate artificially. His plans provide for an illuminating power from each lantern equal to three hundred thousand candles, which is nearly twenty times that of any electric lamp yet manufactured, but is not at all impracticable, as it involves only an increase in electrical volume and pressure and a corresponding increase in the diameter of the carbons. The cost of the tower, lamp and generator for a single light is estimated at fifteen thousand dollars, irrespective of the engine power required to run the latter. Magnificent and original as this conception seems, it has been attempted before, in the infancy of electrical engineering, by a Western experimentalist, who concieved the idea of lighting the city of Cincinnati in a similar manner, by placing enormous lights upon the high ground overlooking the town. This idea was not successful, but possibly the failure was due to the crude electrical engineering of that day, and not to any inherent difficulty.

VARIOUS USES FOR THE ELECTRIC LIGHT.

In the usual rush of business during the fall of the year, the electric light is found to be of great value in evening and night work, particularly in dry-goods establishments, where clear and intense light—one better, in short, than that produced from gas—is desirable in the matching and selection of colors.

Some of the great ocean steamships are provided with the electric light, both for lighting the cabins and steerage and also as a means of preventing collisions. The light makes them visible, it is stated, at the distance of fourteen miles. This provision reminds us that Professor Fleming Jenkin some years ago discovered and patented a new method of lighting the beacons and buoys on the sea coast by electricity, giving a bright, permanent and unmistakable light to guide the mariner, and preserve him from treacherous rocks and shoals. The light is produced by a rapid succession of sparks, due to successive charges and discharges of a condenser situated upon the beacon or buoy. This is charged directly with a voltaic battery, without the intervention of an induction coil. The communication is made by means of submarine wires running from the shore to the beacon or buoy, and can be operated thoroughly by persons on shore. The invention is considered in all respects practicable, and its adoption on the dangerous parts of our coasts would undoubtedly be the means of rendering fewer the dangers of the seas.

About a year ago a number of experiments were made at Metz by a committee of officers in the German army, appointed to investigate the practicability of employing electric light during siege operations, and to suggest any modifications which it might seem expedient to introduce in the apparatus at present in use. Forts Frederic Charles and Alvensleben were illuminated by throwing the electric light upon them, when it was found that at a distance of from two to three kilometers not only buildings but also individual men could be plainly made out. One night the electric apparatus was arranged on the exercising ground outside the Chambiere gate, and the light directed upon a row of targets. Fire was then opened against these latter by a squad of riflemen, and the practice made was nearly as good as that recorded on ordinary occasions when firing by day—a result which was considered exceedingly satisfactory, as a thick mist prevailed at the time, and materially interfered with the action of the light. Altogether, the committee concluded that the electric light may in future be employed with advantage not only in siege operations but also during outpost duty and engagements at night.

Here we must leave the electric light, and devote some space to

THE TELEPHONE.

This instrument is constructed on the principle of the human ear. It consists of an elastic diaphragm, to receive vibrations of air from the human voice or from other sources, so connected with the wires of a battery (or even with wires without a battery) as to communicate the same vibrations in every respect to another membrane or diaphragm situated at a distance. The two diaphragms of a telephone in distant places correspond, in every practical sense, to the two membranes of the human ear, and the connecting wire to the chain of bones between the two membranes. Probably no invention has come more rapidly into popular favor. "It is employed as a means of communication between counting room and factory, merchant's residence and the office, publishing house and printing office, and, in short, wherever oral communication is desired between persons separated by any distance beyond the ordinary reach of the human voice."

THE GERMAN NAME FOR THE TELEPHONE.

In Germany they call the telephone "Farnsprecher," signifying far speaker. The adoption of so short a name, says the *Scientific American*, is a matter of congratulation, because the Germans might easily have found a way of smothering the telephone under some frightfully polysyllabic title. To show how closely the fortunate instrument has escaped this fate, a correspondent in Heidelberg writes that no less than fifty-four names were proposed in German, all of varying degrees of atrocity. Some (we will not inflict the reader with the original titles) signified "mile tongue," "kilometer tongue," "speaking post," "word lightning," "world trumpet," and finally one inventor, collecting all his energies for a grand effort, triumphantly produced "doppelstahlblechzungensprecher." The jaw can be replaced by pressing on the lower molars with the fingers, and guiding the muscles with the thumbs.

THE INVENTOR OF THE TELEPHONE.

Various accounts, as is usual, have been given of the invention of the telephone. An article in a recent number of the Pekin *Gazette*, written by one Chin Hoo, says that Kung Foo Whing, a distinguished philosopher who flourished about the year 976, invented the telephone—which is known in China as "Thumthsein"—in the year 968. It is said that two hundred and ten years ago a book was published in England, in which the author affirmed that "it was not impossible to hear a whisper at a furlong's distance, it having been already done," and that he assured the reader that he had, "by the help of a distended wire, propagated sound to a very considerable distance." The Buffalo *Sentinel*, dated September 10, 1853, contained the following item: "An English paper, the Plymouth *Journal*, announces the discovery of a means of transmitting sounds to a great distance through the medium of water. The instrument by which this is done is called by its inventor a 'telephon' or soundcarrier." These various announcements manifestly do not discredit the statement made by Mr. W. F. Barrett, that the inventor of the electric telephone was Mr. John Cammack. This gentleman says that as early as 1860 Mr. Cammack, while a student in the Royal School of Medicine, Manchester, made and exhibited a telephone containing not only the intermittent current introduced by Philip Reiss, of Hamburg, in 1861, but the principle of continuous current of varying strength used still more recently by Mr. Edison and Professor Graham Bell. There is no evidence, however, that Mr. Cammack had carried out his idea practically like Bell. But what becomes of the claim for Mr. Cammack if it be true that old journals have been found containing an account of a new musical instrument invented by M. Petrina, of Prague, who is stated to have constructed "an instrument with keys which by a galvanic current sets a small iron plate into vibration as soon as the hand leaves the key? Each key produces a different tone, and the tuning and use are similar to that of a pianoforte. A second instrument put at a considerable distance is connected with the other in such a way that the music played on the one resounds from the other." This appears to have been a musical telephone put in practical form long before any now known.

A TELEPHONE SERVICE METER.

At the telephone convention recently held at Niagara Falls, a telephone service meter was exhibited, the invention of Mr. H. L. Bailey, of New York, whereby the time that each subscriber uses the telephone, as well as the number of times it is in use, can be registered by clockwork. If this device realizes the expectations of its inventor, it is probable that instead of the one at present in use, a toll system will be generally adopted, each subscriber paying a nominal amount as rent for the telephone, and so much for every time he uses the instrument, which would doubtless prove more satisfactory both to the company and to the public.

SERMONS BY TELEPHONE.

When the telephone was first introduced it was laughingly said that people need not go to church, but could sit in their own houses and have the sermon and the services wafted to them telephonically. This was done for the first time, we believe, at Lowell, Massachusetts. Twelve persons visited the central telephone office one Sunday morning, on the invitation of Manager Glidden. The office was connected by telephone with the Freewill Baptist Church, an instrument being arranged out of sight behind the pulpit. The organ voluntary rang out clear and sweet upon the ears of the telephone listeners, and the reading and praying—even when spoken in a whisper —were distinguished word for word. Then came the voice of the minister—"We will sing the 428th hymn, omitting the third verse," and after a brief interlude by the organist, the voices of the congregation were heard in pleasing melody. After reading a number of notices the text was announced as a portion of Matthew 16:3: "But can ye not discern the signs of the times?" It was a discourse written evidently for the occasion, and went to establish the truth of the assertion that "science ever has been and must be the safeguard of religion." What science had already accomplished for the world and what religion owed to it were dwelt upon with much force. Before concluding, the minister spoke of some of the wonderful inventions of the day, and made special reference to the phonograph and telephone. During the discourse there was the least possible difficulty in distinguishing the remarks of the preacher when his earnestness in his subject impelled him to emphatic sentences. The moderate tones were all plainly heard, as were also the concluding organ selection as the congregation passed out, and the muffled monotonous tones of the retiring worshippers.

The second experiment of transmitting a sermon by our most noted preacher, created a greater stir. Listeners were not slow to appreciate the novel advantage of listening to a sermon preached by Henry Ward Beecher, without the necessity of crowding into his church.

Owing, says a reporter on the New York Press, to the necessity for concealing the transmitters from the congregation, as well as to the drawback of having but one wire, the sound was not at all times distinct, but was interrupted by inquiries from various points on the circuit. Whenever the preacher thumped his Bible there was a whiz and whirr that was anything but solemn. The music of the choir of the congregation and the soloist were heard plainly all over the circuit. The sermons were rather disconnected, from the fact that the listeners at the instruments were constantly changing, and occasionally the wires would get crossed or the plugs pulled out, so that the discourse would get mixed with messages. The morning sermon ran something like this:

"What can be more pitiful (Hallo! hallo!) than the spectacle recently presented at West Point? (Hallo! Chin—referring to Mr. Chinnock, the electrician—don't cut me off.) How is the young man treated? (There, you've cut me off again.) He was ostracised by his comrades. (Hallo! Beach! Hallo!) Insults were showered upon him. (Put that plug in a little tighter.) He works his way onward. But the detestable prejudice of those who should have been his comrades and associates (Stop calling and listen, will you?) single him out (Brown, be quiet) for persecution, and the brutal (whrr-r-r-r-r!—caused by the preacher pounding the table), cowardly outrage (here a sound like the clashing of cymbals), with accounts of which the newspapers have been teeming for a week, is committed upon him."

As soon as all the listeners got quiet, however, the sermon was heard with distinctness, and when the number of listeners on the circuit was decreased, the sound became much more distinct than when the circuit was open for all. Mr. Chinnock said that with a separate transmitter and a separate wire there would be no trouble whatever in hearing the whole of any service without interruption. The peculiar tone and accent of the preacher were easily recognizable, and the sermon might have been heard by any one of the thirty-five hundred people in communication with the telephone exchange. Mr. Chinnock says that it would be possible for a preacher to stay at home and preach his sermon to a congregation of ten thousand at their homes.

USES OF THE TELEPHONE.

The telephone promises to be of extensive use in

very diverse ways. By means of it, music played in France has been distinctly heard in England.

The telephone is being rapidly introduced into the various military establishments, not only in the capital and its neighborhood, but also everywhere in Germany.

The young Spanish king, now a happy father, being separated from his bride by the rigid court etiquette and public affairs for several days each week, had his private apartments connected with her palace by a telephone, through which the royal lovers communicated without interference or annoyance.

The telephone has lately been successfully used in France to communicate between a vessel being towed and one towing. The wire was carried along one of the hawsers, and the circuit completed through the copper on the bottoms of the ships and the water. Conversation was carried on very distinctly.

Its aid has been secured in Jersey City in connection with the courts. A telegraph wire has been constructed from the Hudson County Court House to the telegraph office in Montgomery Street, and a telephone attached to each end, whereby lawyers can communicate with each other rapidly or between their offices and the court house.

HUMORS OF THE TELEPHONE.

A correspondent thinks that the telephone will soon be utilized on freight trains, so that the conductor can sit by the stove in the caboose and swear at the brakeman, instead of having to go out on the top of the cars in the cold to do it.

Mr. Basingbal (city merchant)—"Most convenient! I can converse with Mrs. B. just as if I was in my own drawing-room. I'll tell her you are here." (Speaks through the telephone.) "Dawdles is here —just come from Paris—looking so well—desires to be," etc., etc. "Now you take it, and you'll hear her voice distinctly." Dawdles—"Weally!" (Dawdles takes it.) The voice—"For goodness' sake, dear, don't bring that insufferable noodle home to dinner!"

The following advertisement appeared in the New York *Herald* "Personal" column at the time that the telephone was first introduced in New York:

"A chance to be married by the Bell speaking telephone will be given to a limited number of couples during the latter part of this month. No charges will be made; satisfactory references required. Applicants should address box 229, *Herald* office."

The ceremony did not, however, for some reason or other, take place, although a marriage by telephone would seem to be fully as appropriate and practicable, so to speak, as one by telegraph.

Magnet writes: "I had often read of the singing-telephone; but I shall never forget the first time I heard one. I was night operator at a small railroad station. Along about four o'clock in the morning, while I was lying on a table, I heard that which seemed to me as some one humming the tune of

"Hold the Fort." As no one was around the depot at that unusual hour of the morning, I came to the conclusion that it was not coming from human lips. So I got up and went outside of the office and listened. As I could hear nothing, I went back in the office, and could still hear the singing, though it soon ceased. After waiting a few moments it commenced again. This time it was "Sweet Bye and Bye." After searching inside and out of the office, I could not tell where the sound came from, and, as I am not the bravest man in the world, I will confess that I began to think of sprites singing in the air. At that instant the armature of the relay on a local wire rattled tremendously, and made a very strange, loud noise. I rushed over to it, and, to my horror, the instrument was singing! Kind reader, imagine my feeling at a lonesome station, at four o'clock in the morning, and, to my knowledge, there had been four men killed within a stone's throw of the office, and the instrument singing hymns! It was more than I could stand. I rushed out of the office, intending to make a home run, when it flashed across my mind that it was the singing telephone."

In Pine Bluff there is a prominent man. There are many prominent men in Pine Bluff, but this one is so very prominent in a certain direction that his name is known along the crowded street and out in the furrowed globe. It is almost unnecessary to call him Colonel C. A man of striking intelligence and profound reading, he has taken up a financial hobby. He knows so well that the United States government should adopt his theory that he would be willing to bet his eternal existence on it He'll stop a man on the street and hammer him with argument, belabor him with deep-set expressions, and kick him with "important information" for hours. One day the Colonel went into M. L. Jones' office, and had just begun to draw himself up for a three-hours' speech, when Mr. Jones remarked:

"By the way, colonel, have you ever seen the telephone work?"

"No; and I don't believe you can hear any better through that thing than you can through a cow's horn."

"I've got one here connected with Colonel Grace's office, and if you'll just put your ear here I'll show you. I'll do the talking—you listen."

The parties took position, Colonel C. incredulously, and Mr. Jones called:

"Colonel Grace, are you there?"

"Yes; is that you, Jones?"

"Yes; how do you feel?"

"I'm about worn out. That —— man C. has been around here this morning boring me to death with his financial business. I guess I'll get rested though after a while."

Colonel C. took his ear away and remarked:

"If he'd only listened to me he would have been smarter in ten minutes more than he ever was before in his life."

Mr. Armstrong, superintendent of the Suburban Telegraph Company of Cincinnati, was on a visit to Chicago at the time that the musical telephone first began to attract attention. On his return he reported that he had made arrangements to test the telephone between the two cities. The music of a brass band at Chicago was to be transmitted over the wires and distinctly heard in Cincinnati. Out of courtesy to the newspaper fraternity, it was announced that none but members of that profession would be admitted to the first trial. When the time came thirty newspaper men were present, pencils and all. It took Mr. Armstrong some time to adjust things properly, but finally sweet sounds were heard. Musical critics, reporters and editors placed their ears close and could not conceal their joy. "I hear the telephone whir," said one; another threw his hat in the air with delight, while the remainder fell upon each other's necks to weep. Presently one of the party said he could distinguish the French horn from the bass drum, another thought the man playing the trombone was blowing too hard to make artistic music, another could count just sixteen pieces in the band while still another counted seventeen. Everybody listened and drank in the delicious strains. Finally the music abruptly stopped. As they all wanted to examine the telephone to its bitter end, Mr. Armstrong lifted the top of the relay box and disclosed a little Swiss music box, which on being wound up struck up: "A Life on the Ocean Wave, A Home on the Rolling Deep." The faces of the astute newspaper men very visibly lengthened as they contemplated what a complete sell had been perpetrated upon them. It is presumed that Mr. Armstrong properly appreciated the fun. He had just graduated in the same school at the hands of Mr. Summers at Chicago a few days before. Some of the reporters felt quite blue over the sell, as a number of them had been studying scientific works on the transmission of sound for weeks, and had several columns of introduction in advance, which was already in type.

THE PHOTOPHONE.

One of the latest marvels in applied science is the discovery by Professor A. Graham Bell and Sumner Taintor of Watertown that " sounds can be produced by the action of a variable light from substances of all kinds, when in the form of thin diaphragms." In other words, a ray of light is substituted for the connecting wire, and sounds at one station are reproduced at another. As is well known the action of the telephone is due to variations in an electric current, caused by a diaphragm set in vibration by the voice, the current thus modified reproducing its variations on a sensitive diaphragm at the other end of the circuit. In the "photophone," as the new invention is called, the changes in the electric current are made during its passage through selenium, a substance heretofore known only as a chemical curiosity, but with the strange property of conducting electricity more easily when exposed to light than when in the dark. A steady light allows a current to pass through an even resistance; a varied light varies the resistance, so that the current is stronger or weaker after passing through the selenium, and its variation are easily turned, in a telephone, into vibrations of sound. Professor Bell and Mr. Taintor have already spoken between points about 600 feet apart, and they believe that the result can be obtained as far as a beam of light can be flashed. The simplest apparatus of many devised consists of a plane mirror of flexible material, as silver microscope glass or mica, which will quiver with vibrations of sound. On this is gathered through a lens a beam of light from any source, success having been found with a kerosene or candle flame. The parallee beam reflected from the plane mirror is thrown to a distant concave mirror and focussed on a piece of selenium, electrically connected with a telephone. The voice throws the plane mirror into vibrations which modify in intensity the ray of light, which rapidly changes the resistance of the distant selenium, this varying the electric current in the telephone as the voice now does directly. Another means of affecting the beam of light is by a disk, preforated with slits, which is rapidly turned, producing in the selenium a continuous musical tone, whose pitch varies with the rapidity of the disk's rotation, a silent motion thus producing a sound. A strange thing is that some substances placed in the beam of light do not cut off the sound. A sheet of hard rubber, for instance, made the beam invisible, but the musical note was still heard. Other experiments suggest the possibility of doing entirely without the electric current in the telephone at the receiving station. Many other substances were substituted for selenium, the affected ray of light focussed upon them, and the musical note was heard without the aid of a telephone or battery. Only carbon and thin glass failed to give a sound.

HATCHING BY ELECTRICITY.

Silk-worms hatched by electricity are now being reared in Italy. The same method is also applied to hens' eggs, and to hastening the germination of seeds.

THEATRICAL THUNDER.

An enterprising citizen of Chicago has invented a process by which real genuine thunder and lightning can be produced by means of an electric battery. The new theory in theatrical thunder is soon to be tried, and the effect produced is said to be startling.

TOOTHACHE CURED BY ELECTRICITY.

Dr. Bouchard, of Paris, says that the toothache may be almost instantly arrested by a constant battery current from ten cells. The positive pole is placed against the jaw, on a level with the painful tooth, and the negative pole to the antero-lateral region, on the same side of the neck.

A SUGGESTION.

There are contrivances for turning gas on and off by electricity, lighting any number of burners at the same instant of time. By connecting this with the burglar-alarm telegraph, the opening of a door or window would set the bells ringing and light all the burners in the house at the same instant.

ELECTRICITY AS A WATER-SHED.

A Frenchman has discovered that electricity applied to a certain small apparatus repels rain. He places the electrical apparatus in his cane, which he holds above his head, when the rain pours off in all directions. The people of the town in which he lives gaze upon him, it is said, with a sort of awe, as he walks in the midst of rain without being wet.

TAMING HORSES BY ELECTRICITY

An English journal says: "Mr. George Laycock, farmer, of Whittington, near Sheffield, was convicted in the penalty of forty shillings and costs, by the Sheffield stipendary magistrate, for cruelty to a mare, which he was taming by electricity at a public sporting ground. Horse-taming by electricity in Yorkshire has, it is said, been freely practiced of late, and the prosecution therefore excited considerable interest."

ELECTRICITY AND RELIGION.

At the Moody and Sankey meetings in New York, the several halls of the Hippodrome were connected by telegraph, and when the director, sitting on the platform immediately behind Mr. Moody, wished the gas turned on, the doors or ventilators opened or closed, or the like, he did it by that agency. Small electric bells were also arranged around the building, on which orders to the door-keepers, ushers, etc., were given. Everything worked like clock-work.

ELECTRICAL AIR AS A TRANSMITTER.

It has long been known that telegraphic messages could be transmitted without the use of wires, and many years since signals were sent across the Bristol Channel by the use of the water as the conducting medium; but in that case the water through which the signals passed was inclosed in a tube, so that it was, in truth, only the substitution of a wire of water, if the term can be used, for the metallic wire usually employed. Professor Loomis now proposes to go further; he claims to have discovered a mode of transmitting messages by electrical air currents, and is seeking an opportunity for making experiments on the summit of Mont Blanc.

MAPS BY TELEGRAPH.

A member of the Parisian Academy of Science has devised a method whereby exact maps and diagrams may be transmitted by telegraph. A numerally-graduated semi-circular plate of glass is laid by the telegrapher over the map to be transmitted, and a pencil of mica, attached to a pivoted strip of metal, also divided into numbers, allowed to move over the plate. Looking through a fixed eye-piece, the operator traces out his map on the glass with the adjustable mica pencil, and, noticing the numbers successively touched on the plate and on the moving metal arm, telegraphs them to his correspondent, who, by means of an exactly similar apparatus, is thereby enabled to trace out an exactly similar map.

MAGNETIC MAGIC WRITING.

In Bristol, Professor Thompson recently made an interesting experiment, which can be used as a secret or magic writer, and reminds us of the magic inks which appear by heat and disappear again by cooling. He took a very thin sheet of hardened steel, and made invisible letters on it by means of the point of an iron bar strongly magnetized by means of a surrounding coil and battery; he found that all the places touched had become permanently magnetic to such a degree that when fine iron-filings were placed upon it, and then the plate turned over to make them fall off again, the iron-filings remained on the spot touched with the magnet, and thus made the writing visible. The writing may be rubbed out by brushing the filings away, but reappears any time afterward when the filings are again applied.

ELECTRIC DRIVING POWER.

The New York correspondent of the Boston *Journal* describes a new invention for displacing steam by electricity, and says that lathes, planing machines, and other mechanical arrangements are driven by this power. To run an engine of twenty-horse power by this invention would require only a space of three feet long two feet wide and two feet high. The cost per day would be thirty-five cents. On a steamship no coal would be required, and the space now used for coal and machinery could be used for cargo. The stubborn resistance of electricity to mechanical use heretofore has, it is believed, been overcome. A continuous battery has been secured, and other difficulties removed, principally through the coil of the magnet. If the invention works as well on a large scale as it does on the machinery to which it is now applied, steamships will soon ply the ocean under the new propelling power. The whole thing, mighty enough to carry a Cunarder to Liverpool, can, he adds, be secured in a small trunk.

ELECTRICITY IN MANAGING REFRACTORY HORSES.

The French papers tell of a wonderful invention which will enable the feeblest among us to "witch the world with noble coachmanship." The horse of the future is not to be driven by ordinary reins, but by electricity combined with them. The coachman is to have under his seat an electro-magnetic apparatus, which he works by means of a little handle. One wire is carried through the rein to the bit, and another to the crupper, so that a current once set up goes the entire length of the animal along the spine. A sudden shock will, we are gravely assured, stop the most violent runaway or the most obstinate jibber. The creature, however strong and however vicious, is "at once transformed into a sort of inoffensive horse of wood, with the feet firmly nailed to the ground." Curiously enough, the very opposite result may be produced by a succession of small shocks. Under the influence of these the veriest screw can be suddenly endowed with a vigor and fire indescribable, and even the Rosinante of Don Quixote would gallop like a racer. What is the effect upon the condition of the horse is not stated, but the *Siecle* finds itself able to congratulate M. Fancher upon "an invention equally original and salutary," and one which places in the hands even of an infant a power over the horse which is as sovereign as it is invisible.

ENGRAVING BY ELECTRICITY.

A novel apparatus for engraving by electricity was exhibited in the machinery department of the French exhibition. A metal plate, with some object drawn upon it with a special ink, is slowly rotated with its face vertical; and several other similar plates, but of decreasing smallness and with correspondingly diminished speed, are also slowly rotated by appropriate mechanism. On these plates it is intended the object delineated on the first plate shall be engraved on different scales of magnitude; and this is accomplished by applying a diamond cutting-point to the face of each plate, which is pressed against it through the agency of an electrical current whenever a blunt point presented to the first plate encounters the ink, but is withdrawn at other times. The point presented to the first plate is a "feeler," which determines by electrical agency whether there is ink beneath it or not. If there is, the diamond points opposite to all the other plates are pressed in, and if there is not, they are withdrawn and prevented from cutting. The "feeler" and the diamond burins must all follow a spiral track.

DIAGRAMS OF TARGETS OVER THE WIRE.

This feat, which at first sight seems an almost incredible one, looks more simple when it is suggested that there be prepared in the editorial sanctum, beforehand, two similar sheets, each the size of the targets to be used and ruled very closely in two directions, so that the lines intersect. Then number every line on the margin. The reporter uses one sheet, and by saturating it with oil it will, if thin, become sufficiently transparent to enable him to trace with lead pencil the marks on the targets. What easier then than to send by telegraph the intersections, which may be made frequent enough to locate so closely as to answer all practical purposes?

ELECTRIC COMBS AND BRUSHES.

In an old number of the *Scientific American* we find the following, which is interesting inasmuch as the suggestion it makes has been acted upon in the production of a hair-brush now freely advertised.

"The manufacture and sale of hair restoratives has always been a favorite with a certain class of public benefactors, whose disinterested labors have resulted in the foundation of many a fortune. We lately came across the specifications of an old English patent which will, perhaps, be interesting at a time like the present, when alcohol and bear's grease command such fabulous prices.

"This patent was for 'an apparatus for improving and restoring the human hair,' introducing a new feature in this line. By the plan of this inventor combs and brushes are to be constructed of different metals, so that when in use electric currents are given off; 'thereby the skin is caused to be stimulated, and a healthy action ensues, restoring the hair to its original color, and generally improving its appearance.' The same effect may be produced by having the articles formed partly of metal having batteries connected therewith when in use. As the patent claim long since expired, the above method is open to any enterprising individual wishing to experiment."

NEW USES FOR THE SUN'S RAYS.

The thermo-electric battery is exciting the imaginations of men of science, causing them to call up wonderful visions of a future when much of the work of the world shall be done by sunshine. Like windmills, thermo-electric batteries might be erected all over the country, finally converting into mechanical force, and thus into money, gleams of sunshine, which would be to them as wind to the sails of a mill. What stores of fabulous wealth are, as far as our earth is concerned, constantly wasted by the nonretention of the solar rays poured out upon the Desert of Sahara! Nature here refuses to use her wonderful radiation net, for we cannot cover the desert sands with trees, and man is left alone to try his skill in retaining solar energy. Hitherto helpless, we need not be so much longer, and the force of a Sahara sun may be carried through wires to Cairo, and thence irrigate the desert; or possibly, if need be, it could pulsate under our streets and be made to burn in Greenland.

THE OCEAN A SOURCE OF ELECTRICITY.

An important experiment was made by M. Duchemin, of Paris, during a holiday at the seaside. He made a small cork buoy, and fixed to it a disc of charcoal containing a small plate of zinc. He then threw the buoy into the sea, and connected it with copper wires to an electric alarum on the shore. The alarum instantly began to ring, and went on ringing, and it is added that sparks may be drawn between the two ends of the wires. Thus the ocean seems to be a powerful and inexhaustible source of electricity, and the small experiment of M. Duchemin may lead to most important results.

ELECTRICITY AS AN EXECUTIONER.

It has been proposed to substitute a method less clumsy than those now obtaining for the execution of criminals, and the adoption of electricity for this purpose has enthusiastic advocates in Germany, as well as in France, as witness the following imposing description of a method proposed by a German writer: "In a dark room, draped with black, and which is lighted only by a single torch—the chamber of execution—there shall stand an iron image of Justice with her scales and sword. Stern Justice is popularly supposed to have no bowels, but the German goddess will carry a powerful battery in her inside; and this battery will be connected with an arm-chair—the seat of death. In front of the chair will stand the judge's tribunal, and only the judge, jury, and other officers will be present with the criminal during the ceremony of the execution. This will consist in the judge reading the story of the crime committed by the prisoner, who will be rigidly manacled to the aforesaid arm-chair, and when this is done, the judge will break his rod of office, and toss it into one of the scale-pans of justice, at the same time extinguishing the solitary torch. The descent of the pan will complete the electric circuit, and shock the wretch into the next world."

It is also suggested to utilize the electric fluid in killing animals. A battery and coil would be far more effective and far less cruel tools than the pole-axe or the sticking-knife We suppose the angler would consider his occupation gone if he had to fish with an electric line and a torpor-producing bait; yet the whaler has a notion that he can catch his monsters upon an analogous plan. A London firm have obtained a patent for a method, startling to "old salts" in its originality, for catching whales by means of electricity. By their plan every whaleboat is provided with a galvanic battery. Wires from opposite poles run down to the points of each set of harpoons. When the whale is near, two harpoons are thrown as nearly simultaneously as possible, which when embedded in the flesh of the monster, complete the circuit. The charge is expected to be sufficiently powerful to paralyze the animal, so that the small boat may advance and dispatch him at leisure.

ELECTRIC CLOCKS.

A citizen of Burlington, Vermont, has invented a clock that runs by electricity, and never requires winding. It has only three wheels, no weights or springs, and it is claimed that it has little friction, is not affected by heat, cold, dampness or jarring. A single clock and battery can be connected with any number of dials and indicators in the same building, or even along the whole line of a railway.

A magnetic clock, invented by Daniel Drawbaugh, of Milltown, Cumberland county, Pennsylvania, is sufficiently remarkable to be wor. description. The magnetism of the earth, an inexhaustible source of power, is made to oscillate the pendulum, and the simplicity of all the works gives an assurance of the least possible friction. At a certain point the movements of the pendulum itself shut off magnetic connection with the earth, and at another point restore the connection, thus securing conditions necessary to produce its oscillations. The works are so ingenious and simple that it is no wild assertion to make that, were it not for the unavoidable wearing out caused by even the small amount of friction, the clock would run as long as the solid earth endures. This clock was hung against the board partition, with all its works exposed, subject to the jarrings of machinery and obstructions from dust settling on it, for years, yet it ran continuously and uniformly, with only slight reported variations, as tested by transit observations at noon.

STEAM AND ELECTRICITY

Mr. W. H. Bailey, an English inventor, has proposed a new system of sea telegraphy, by means of which vessels can communicate in foggy weather, or when a considerable distance apart. It consists simply in the adaptation of the Morse code of signals to a steam whistle. The message is read by ear, the whistle, worked by a hand lever, giving forth long and short sounds corresponding to the long and short lines of the Morse system. According to experiments made by the inventor, it is believed that a twelve inch whistle can be heard at a distance of six miles, and that two vessels passing within hearing could converse at the rate of twelve hundred words an hour. The advantages of such a system in foggy weather are evident.

It would startle many people, who happened to see a locomotive blowing off steam at a railway station, if they were told that there is electricity enough generated in the discharge of steam to blow the whole train to atoms, if, instead of being dissipated, it were collected. The fact was first accidentally noticed by an English engineer, who perceived sparks, which proved to be electrical, among the escaping steam. The discovery was confirmed by the construction of a hydro-electrical machine in the shape of a boiler set on glass legs. The steam, as it rushes out of the escape valve, is received on a series of metallic points by which it is gathered and accumulated in the conductor, as in an ordinary electrical machine, in which the electricity is generated by the friction of a glass-plate or cylinder. Will engineers ever come to appreciate the fact that every locomotive, or tug, or steamer carries the means of lighting itself far better and more cheaply than by any lamp?

THE EDISON ELECTRIC LOCOMOTIVE.

The Edison electric locomotive is about the size of an ordinary hand-car which railroad laborers propel along the track, and consists simply of one of Edison's generators on wheels. When this apparatus is intended to generate electricity, the armature is turned with great rapidity by two powerful magnets, and takes from them a quantity of magnetism or electricity, which is used for any purpose for which it may be needed. A steam engine of at least five-horse power is needed to turn the armature of one of these generators. In the locomotive the generator receives instead of generating electricity, and the armature turns with great rapidity as the current passes through it. It is like winding up silk on one bobbin and unwinding it on another. In running the locomotive, therefore, two generators are used, one stationary in the engine-house, worked by steam and generating the current, and the other on the locomotive receiving motion from the current. The armature on the locomotive is geared to the driving-wheels, so that it makes four revolutions to one of the driving-wheels. It is as if the stationary engine wound up a spring in one generator to be let loose and impart motion to another. Electric motors are plenty as blackberries, and toy locomotives going by electricity have been made to run around a table. Dr. Siemen, of Berlin, and Edison are the first to construct locomotives of any size. The problem has always been to get the electricity to the engine without having to carry along the whole generating apparatus on the train. The new plan is to make the track carry the current. It makes no difference whether the locomotive is standing still or going at the rate of fifty miles an hour, so far as receiving the current through the rails is concerned. The current reaches the locomotive wherever it may be found on the track, and entering by the wheels reaches the armature and sets it revolving.

ELECTRICITY AIDING WEARY CASH GIRLS.

An enterprising dry goods firm in this city have recently tried the experiment of running their cash system by electricity, and with excellent results. Previous to the introduction of electricity, on Saturdays, particularly in the afternoon, the din and confusion, and the incessant call of "cash!" "cash!" "cash!" by the saleswomen and salesmen were absolutely deafening. So the two Ehrich brothers put their heads together to invent something that would call the cash girls without so much noise. "I suggested bells," says Mr. Ehrich, in telling the story, but Louis said: "'No, that would be as bad as the cash calls.' One day he came to me and said, excitedly: 'William, I've found it. Electricity is the thing.' I declare I thought Louis had gone crazy. 'Found what?' said I. 'What is electricity the thing for?' 'Our cash girls,' he replied. 'In the name of conscience, Louis,' said I, 'what are you going to put electricity on our cash girls for? I don't see that anything is the matter with them.' Then Louis began to laugh. He explained that he meant to apply electricity to call them, instead of the cash call used in all the stores in the city from A. T. Stewart's to ours. Come and see his invention." And he led the way to the register in the center of the store under the main staircase, where there are thirty or more little circular silver-plated drops, labelled "hosiery," "buttons," "millinery," and so on, with numbers also to correspond with the sections. Every now and then, as if by magic, down dropped one of the little silver plates. A young man standing by the side of the register instantly spoke, "hosiery," or "trimmings," 1, 2 or 3, as the case might be, and as soon as he thus announced the department and number, off started the head girl in the line of cash girls seated on the other side of the register. In the mean time others came up as fast as the first departed and took their seats in the line. There was no confusion, no hurry, not a call throughout the large and busy establishment although dollars and parcels by the hundred were passing over the counters. Near each of the counters are little cord-like straps running back of the saleswomen, that they pull whenever a purchase is made and a sale completed, and which are connected with electrical wires running under the floors and joined to the drops at the register.

CONCLUSION.

We must now bid our reader farewell, trusting that he will have enjoyed the variety of entertainment and appreciated the instructive matter presented in the foregoing pages. We also trust that he will find them useful for future reference and companionable in solitary hours to come. Of his charity, we ask him to take them as they are and as they profess to be. Then, we modestly assure ourselves, he cannot be disappointed, and humbly believe that he will be abundantly satisfied.

www.ingramcontent.com/pod-product-compliance
Lightning Source LLC
Chambersburg PA
CBHW021410230426
43666CB00006B/695